Magnetic Microwires

Alexander Chizhik | **Julian Gonzalez**

Magnetic Microwires
A Magneto-Optical Study

PAN STANFORD PUBLISHING

Published by

Pan Stanford Publishing Pte. Ltd.
Penthouse Level, Suntec Tower 3
8 Temasek Boulevard
Singapore 038988

Email: editorial@panstanford.com
Web: www.panstanford.com

British Library Cataloguing-in-Publication Data
A catalogue record for this book is available from the British Library.

Magnetic Microwires: A Magneto-Optical Study

Copyright © 2014 Pan Stanford Publishing Pte. Ltd.

ISBN 978-981-4411-25-7 (Hardcover)
ISBN 978-981-4411-26-4 (eBook)

Printed in the USA

Contents

Preface

The idea of studying magnetic wires using the magneto-optic technique appeared in the late nineties of the 20th century as a response to the discovery of the giant magneto-impedance (GMI) effect—one of the most promising effects observed in magnetic wires. This idea looked very attractive because of the following reasons. The systematic magneto-optical investigation of the no-plane objects had not been performed ever before. Usually the magneto-optical technique is used to study plane objects, and the present task of the magneto-optical investigation of cylindrically shaped samples attracted me by unusual and original experimental configuration. By that time I had acquired a good enough background as an experimentalist in magneto-optics, obtained in the well-known magneto-optical school belonging to the B. Verkin Institute for Low Temperature Physics and Engineering in Kharkov, Ukraine, and with 20 years of joint work with such great scientists as academicians V. V. Eremenko, N. F. Kharchenko, and S. L. Gnatchenko, who have taught me to always search the original promising scientific tasks.

Another motivation was determined by the "surface" nature of the GMI effect—the penetration depth of the ac current changes in the presence of dc applied magnetic field. As it is known, the magneto-optical Kerr effect, which is usually used for the investigation of non-transparent objects, is also the "surface" effect—the light reflected from the sample contains the information about the magnetic behaviour in the thin near surface layer. In this way, the task of obtaining a deep understanding of the surface magnetization processes that affect the skin effect in magnetic conductors has found a suitable powerful method.

It is necessary to remark that the great variety of the chemical composition, sizes, and geometric ratios (metallic nucleus)/(glass shell) promises a wide range of interesting magnetic effects to be studied.

The magneto-optical study of magnetic wires was initiated by Profs. J. M. Blanco and J. Gonzalez, whom I appreciate very much for their kind and farsighted invitation of collaboration. The main part of the magneto-optical investigations was performed in the Laboratory of Magnetism of the University of Basque Country, Spain, with which I was associated during the past 10 years of fruitful activity. It is necessary to note the creative and open atmosphere in the lab formed by Prof. A. Zhukov, Dr. V. Zhukova and Dr. M. Ipatov.

Also, a significant number of the magneto-optical images presented in the book were obtained in the Laboratory of Magnetism of the University of Bialystok, Poland, under the direction of Prof. A. Maziewski. These images are the result of my collaboration with Prof. A. Stupakiewicz, a great specialist in magneto-optics.

An essential objective of our study was to perform a theoretical analysis to establish the depth of the understanding of the discovered effects. I have performed this analysis in association with Prof. K. Kulakowski, Dr. P. Gawronski from Krakow, Poland, and Prof. V. Zablotskii from Prague, Czech Republic.

Almost all the glass-covered microwires studied in our laboratory have been supplied by a small but very powerful company, Tamag Iberica. The company has a wide international collaboration from Moldova to Japan and has always provided us the best experimental material.

Two directions were prioritized in our investigations. First, I sought the discovery of new effects, and I succeeded in it: circular bistability, visualization of the giant Barkhausen jump, correlation between the GMI effect and surface magnetization, etc. The discovery of these phenomena has led us to new level of the understanding of the physical processes that take place in magnetic wires. The significant experimental success was conditioned by the realization of the original experimental configuration named "magneto-optical Kerr effect without external magnetic field": The studies have been performed only in the presence of a circular magnetic field produced by the electric current flowing along the microwire.

The other fundamental task was the optimization of the magnetic, electrical and mechanical properties of the wires taking into account the technological application. The performance of the systematic study of a wide series of wires of different compositions and geometric ratios permitted us to determine the methodical features that help to predicate and control the key parameters of microwires.

The contents of this book reflect this duality of our investigation. The book describes the basic physical effects and the chapters are devoted to the detailed elucidation of the observed features.

Directing the attention to the glass-covered wires of the micrometer scale, I have advanced to the nano-area. Now it is clear that the cylindrical symmetry and the specific stress distribution induced by the glass covering are the key parameters that provide the competitive technological application of microwires. The idea is to move to the nanoscale while maintaining the achieved optimal parameters of microwires. The first steps of this shift to the "nano" range are presented in this book as an example of the versatility of the basic effects initially discovered for the microscale.

I hope this book will serve as a landmark that encourages us to apply the experimental and theoretical efforts to the discovery of new surprising effects in magnetic wires.

I would like to express special thanks to my wife, Marina, daughter, Daria, and mother, Zoia, for their continuous support and their interest in my work.

Alexander Chizhik

Chapter 1

Kerr Effect as Method of Investigation of Magnetization Reversal in Magnetic Wires

1.1 Introduction

Optical effects that demonstrate the influence of the magnetization on the state of light are known as magneto-optical effects. Two main magneto-optical phenomena are usually distinguished: the Faraday effect and the Kerr effect. The Faraday effect is characterized by the rotation of the polarization plane and the change of the ellipticity when linearly polarized light passes through the magnetized medium. The Kerr effect [1] consists of the rotation of the plane of polarization and the change of the intensity of the polarized light reflected from a surface of the magnetized matter. There are three magneto-optical Kerr effects depending on the mutual arrangement of the magnetization and the plane of incidence of the scattered light—polar, longitudinal and transverse. Experimental investigations have shown that the rotation of the polarization or the change of light intensity (depending on

Magnetic Microwires: A Magneto-Optical Study
Alexander Chizhik and Julian Gonzalez
Copyright © 2014 Pan Stanford Publishing Pte. Ltd.
ISBN 978-981-4411-25-7 (Hardcover), 978-981-4411-26-4 (eBook)
www.panstanford.com

the magneto-optical effect configuration) is proportional to the projection of magnetization on one of the three orthogonal axes [2, 3]. Based on this relation, Kerr effects have been used to get the images of magnetic domains or to obtain magnetization loops.

Usually the magneto-optical Kerr effect (MOKE) has been applied to study samples with plane surface. Recently, we have used the Kerr effect to investigate wire shaped specimens. It is remarkable to note that the study of the magnetic properties of amorphous wires is a topic of great interest because of their outstanding properties, such as single and large Barkhausen jump or Giant Magnetoimpedance effect [4, 5]. It is known that the domain structure of the amorphous wire consists of an axially magnetized inner core and the outer shell, in which a "maze" or a "bamboo" structure can be observed [6].

Taking into account that the Giant Magnet impedance effect is mainly a "surface" effect, the investigation of magnetic structure in the outer shell becomes a special meaning. Earlier, the Kerr effect has been used only to get images of the magnetic domains in the surface areas of the wires [7]. In this case, the information about domain structure can be obtained, but details of the magnetization reversal process are not displayed. Now we use the Kerr effect loop tracer to study the magnetization reversal of the wire surface [8]. Therefore, there are three lines of the MOKE study of the composite microwires: magneto-optical magnetometry (loop tracer), MOKE polarizing microscopy, and MOKE-modified Sixtus–Tonks method to study the surface domain wall dynamics. The main idea of the experiments is MOKE without external magnetic field: the studies have been performed in the presence of the circular magnetic field that was produced by electric current flowing along the microwire.

1.2 MOKE Magnetometer (Loop Tracer)

The scheme of the experimental setup of a magneto-optical magnetometer is presented in Fig. 1.1. A polarized light of He–Ne laser was reflected from the wire to the detector. When the longitudinal Kerr effect was used, the rotation of the angle of the light polarization was proportional to the magnetization, which was parallel to the plane of the light. When the transverse Kerr effect was used, the intensity of the reflected light was proportional to the magnetization, which was perpendicular to the plane of the light. An electric current flowing along the wire produced a circular magnetic field. An axial magnetic field was produced by the pair of Helmgolz coils.

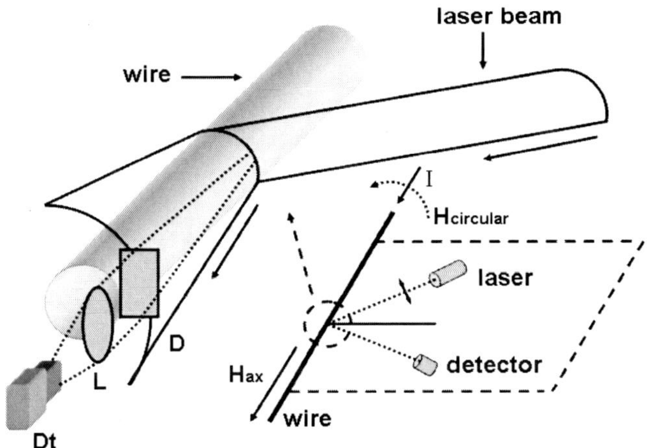

Figure 1.1   Scheme of the Kerr effect setup. D, diaphragm; L, lens; Dt, detector.

The light reflected from the cylindrical surface of the wire forms the conic surface (Fig. 1.1). To avoid the distortion of the magneto-optical signal related to the reflection from the non-plane surface, the part of the light that corresponded to a small area of the wire surface was cut by the diaphragm and the lenses. In consequence, the curvature of this area is estimated to be about 1°.

1.3   MOKE Polarizing Microscopy

Two configurations of the MOKE microscope (Fig. 1.2)— longitudinal (L-MOKE) and polar (P-MOKE) (Fig. 1.3)—have been used during the performed experiments. L-MOKE is manly traditional configuration. We have discovered the domain structure in glass-covered microwires using namely this configuration. The images of the domain structures obtained by L-MOKE microscopy show the difference in plane projection of the magnetization. The special peculiarity of the microwires is the non-planar surface. For the MOKE magnetometry experiments, this peculiarity presents some disadvantage: the light reflected from the cylindrical surface of the microwire forms the conic surface. Normally this problem could be resolved if the part of light reflected from the microwire, which corresponds to a small area of the microwire surface, was cut by the diaphragm [9]. This disadvantage caused by the non-planar surface transforms to an advantage for the case of the P-MOKE microscopy.

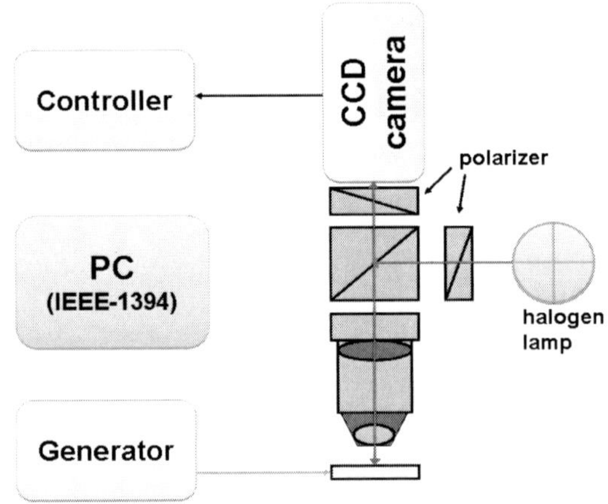

Figure 1.2   Schematic of MOKE microscopy.

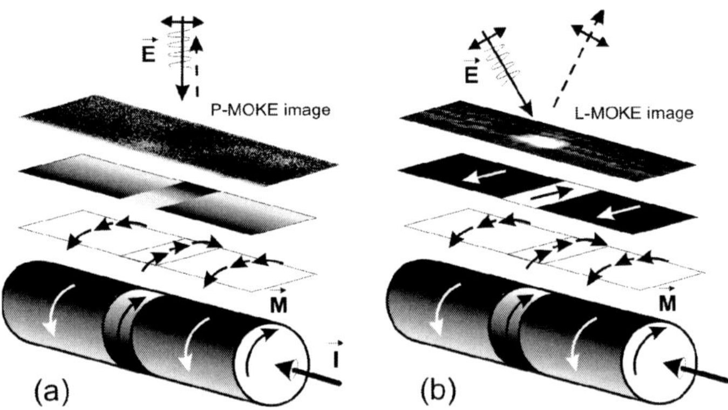

Figure 1.3   Schematic configuration of the circular domain observation in microwire using (a) polar and (b) longitudinal magneto-optical Kerr geometry.

The black–white contrast in the images obtained for this configuration depends on the perpendicular projections of the magnetization. Therefore, the circular magnetic domains could be observed in P-MOKE configuration, but the obtained images are different in comparison with the L-MOKE configuration. The MOKE technique could be applied to glass-covered microwire of a wide range of compositions and diameters. It is not needed to remove the glass cover to perform MOKE experiments. The relevance of this method is determined by that the P-MOKE and L-MOKE configurations of Kerr microscopy are complementary tools, which permits one to know the directions of the magnetization in non-plane surface of studied microwires. The combination of the fluxmetric method, MOKE magnetometry and MOKE microscopy provide complex information about the magnetic structure in different parts of the microwire.

As an example of MOKE microscopy application, Fig. 1.4 presents the images of surface domain structure obtained for P-MOKE and L-MOKE configurations (composition $Co_{67}Fe_{3.85}Ni_{1.45}B_{11.5}Si_{14.5}Mo_{1.7}$, metallic nucleus radius 11.2 µm, glass coating thickness 3 µm). In this figure, we observe the black–white contrast on the gray background. This specific distribution of the colors is caused by the specific conditions of the experiment: in the first stage of the experiment, the magnetic system of the microwire was saturated by the axially directed DC magnetic field. The black–white stripes in this figure are the images of the circularly magnetized surface domains. The black color means the "down" direction of the magnetization and the white color means the "up" direction in the circular domains.

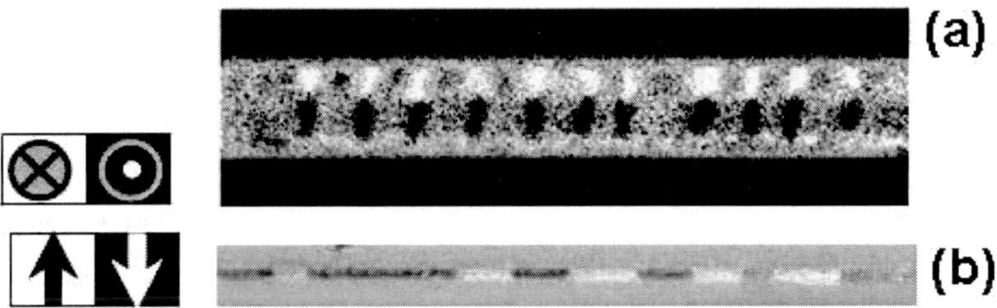

Figure 1.4 Images of surface domain structure obtained for P-MOKE and L-MOKE configurations.

1.4 MOKE-Modified Sixtus–Tonks Method

Formerly, induction methods were used to study the domain wall dynamics. The first one was introduced by Sixtus and Tonks [10] in 1932. Very simple experiments consist of a primary coil to produce the homogeneous field in which the domain wall propagates (Fig. 1.5). The nucleation coil is placed at the end of the wire in order to start the domain wall propagation along the wire. Two pickup coils are connected to the oscilloscope and two sharp maxima appear when the single domain wall passes through the pickup coil. The domain wall velocity is given as the ratio of the distance between the pickup coils and the time between the maxima ($v = L/t$). Such a simple experiment has been used to study the domain wall dynamics of a thin microwire [11]. However, when the diameter of a microwire decreases, the signal from the pickup coil is too small to be detected successfully. Moreover, the Sixtus–Tonks experiment cannot be used to study the domain wall dynamics of very fast domain walls since one should take into account also the relaxation time of the pickup coils. This can be, in case of very fast domain, much longer than the proper maximum induced from the domain wall propagation. Therefore, the Sixtus–Tonks method could not be used to study the fast domain wall dynamics in very thin microwires.

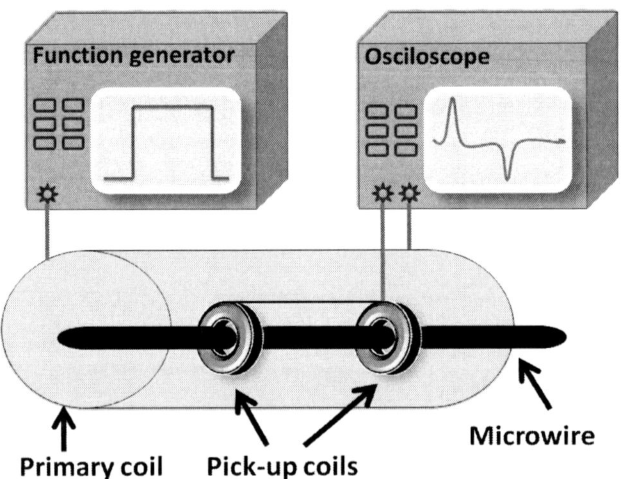

Figure 1.5 Sixtus–Tonks experiment for study the domain wall dynamics in elongated ferromagnetic structures.

Naturally, the idea to combine the Kerr effect (which allows us to study the fast magnetization processes in small structures) with the Sixtus–Tonks experiments (which allow us to determine the domain wall velocity unequivocally and so to study the domain wall dynamics) arises.

Here we present (Fig. 1.6) the Kerr effect–based Sixtus–Tonks experiments to study the single domain wall dynamics in glass-coated microwires. Instead of the pickup coil, the reflection of the broken laser beam from the microwire surface is used. This method allows us to study the propagation of small and fast domain walls [12].

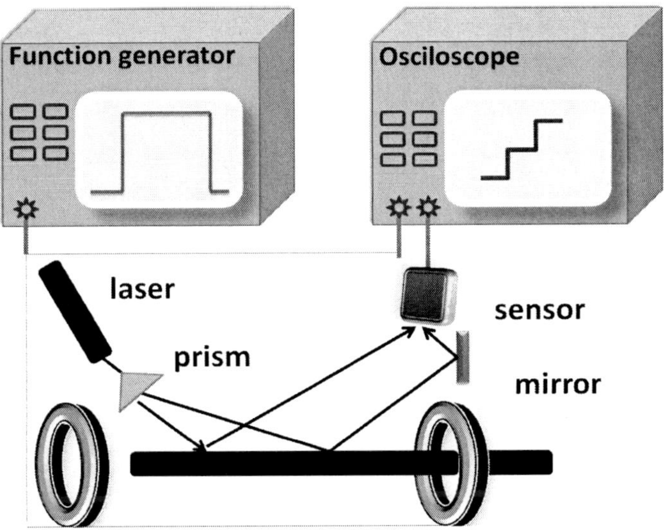

Figure 1.6 Kerr effect–based Sixtus–Tonks experiment for the study of the domain wall dynamics in elongated ferromagnetic structures.

The system consists of primary coil, which is fed by the square shape signal in order to produce the homogeneous field in which the domain wall in microwire propagates. The microwire is placed coaxially in the center of the primary coil with one end outside in order to avoid the two domain walls propagation (from both ends of microwire). Instead of using two lasers, single laser with a prism placed in front is used to get two laser beams. After the reflection on the microwire surface, the two beams are focused with the mirror into the single optical sensor. The output of the sensor is connected to the oscilloscope input in order to detect the time of the domain wall propagation. Two points of reflection are separated by 5 cm.

The Sixtus–Tonks method was adjusted to drive and study the circular surface domain wall (DW). The main idea of this adaptation is the MOKE without external magnetic field—a pulse of electric current flowing through the microwire was used to produce pulsed circular magnetic field. The transverse configuration of MOKE has been used basically in our experiments, when the intensity of the reflected light is proportional to the magnetization, which is perpendicular to the plane of the light (circular projection of the magnetization). A DC axial magnetic field was applied as an external parameter.

The single domain wall motion along the microwire was registered as two successive jumps of the MOKE signal (Fig. 1.7). This means that the single DW passed successively through two laser spots. To determine the velocity of the surface DW, the time duration between two peaks of two derivatives was used.

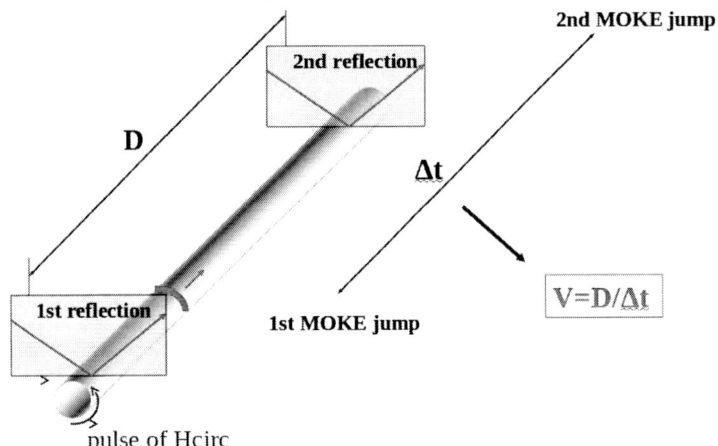

Figure 1.7   Schematic picture of MOKE-modified Sixtus–Tonks method. Inset shows MOKE signal jump related to domain wall motion along the laser spot (black line) and its derivative (red line).

References

1. Kerr J (1887), *Philos. Mag.*, **3**, 321.

2. Sokolov AV (1967), *Optical Properties of Metals* (Elsevier, New York, USA).

3. Hubert A and Schäfer R (1998), *Magnetic Domains* (Springer-Verlag, New York, USA).

4. Beach RS and Berkowitz AE (1994), *Appl. Phys. Lett.*, **64,** 3652.

5. Panina LV and Mori K (1994), *Appl. Phys. Lett.*, **65,** 1189.

6. Mohri K, Humphry FB, Kawashima K, Kimura K, and Mizutani M (1990), *IEEE Trans. Mag.*, **26,** 1789.

7. Takajo M, Yamasaki J, and Humphrey FB (1993), *IEEE Trans. Magn.*, **29,** 3484.

8. Chizhik A, Zhukov A, Blanco JM, and Gonzalez J (2001), *Phys. B*, **299**, 314.

9. Chizhik A, Zhukov A, Blanco JM, and Gonzalez J (2002), *Phys. Status Solidi (A)*, **189**, 625.

10. Sixtus KJ and Tonks L (1932), *Phys. Rev.*, **39**, 357.

11. Varga R, Zhukov A, Blanco JM, Ipatov M, Zhukova V, Gonzalez J, and Vojtaník P (2006), *Phys. Rev. B*, **74**, 212405.

12. Chizhik A, Varga R, Zhukov A, Gonzalez J, and Blanco JM (2008), *J. Appl. Phys.*, **103**, 07E70.

Chapter 2

Cold-Drawn Fe-Rich Amorphous Wire

2.1 Introduction

Cold-drawn amorphous wires have attracted great interest because of their outstanding properties such as magnetoimpedance, stress impedance, magnetic bistability [1–6], which makes these materials as potential candidates to be used in sophisticated applications for sensing devices. We have discovered high value of magnetoimpedance phenomenon, about 60% real component and about 300% imaginary component, in Fe-rich cold-drawn wires [7] that is larger than it was observed earlier [5]. The studied wires demonstrate the field-tension annealing induced bistability and magnetoimpedance effect, but other interesting effects have been also discovered: during the last few years, we have found the existence of helical magnetic structure in Fe-rich cold-drawn wires, and such effects as two Curie points magnetic behavior and quasi-bistability effect. This chapter is devoted to the investigation of the process of the formation of

Magnetic Microwires: A Magneto-Optical Study
Alexander Chizhik and Julian Gonzalez
Copyright © 2014 Pan Stanford Publishing Pte. Ltd.
ISBN 978-981-4411-25-7 (Hardcover), 978-981-4411-26-4 (eBook)
www.panstanford.com

helical magnetic structure, which was not performed earlier in these wires [8–11]. Using the MOKE technique, we pay attention to the process of magnetic transformation that occurs in the surface area of the wire. The study of this type is very important for the cold-drawn wires taking into account the strong transformation of the surface of the wire during the cold-drawing process.

2.2 Two Magnetic Phases in Cold-Drawn Fe-Rich Amorphous Wire

Amorphous wires of nominal composition $Fe_{77.5}B_{15}Si_{7.5}$ and diameter 0.05 mm, obtained by in-rotating-water-quenching technique + cold-drawing process, were provided by Goodfellow Company, UK. The wires were submitted to thermal treatment by using a current annealing technique (210 mA for 45 min). Conventional hysteresis loops were measured by fluxmetric method in an AC axial magnetic field of 50 Hz. A pair of Helmholtz coils has been employed to produce an axial magnetic field. DC axial and DC circular bias magnetic fields have been applied during the magneto-optical and fluxmetric experiments, respectively.

Figure 2.1 presents the bulk hysteresis loops measured by the fluxmetric method with DC electric current as a parameter. The magnetization reversal consists of two successive jumps of magnetization. The shape of the hysteresis loop changes with the DC electric current (i.e., DC circular magnetic field). In particular, the decrease of the switching field for the second jump takes place.

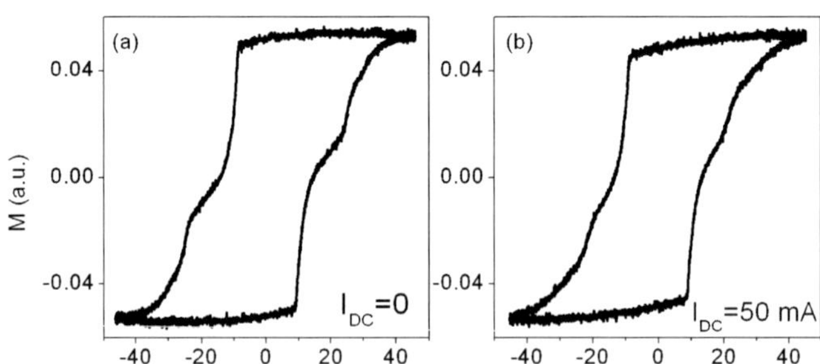

Figure 2.1 Bulk hysteresis loops with DC electric current as a parameter.

Figure 2.2 presents the AC circular magnetic field (created by an AC electric current) dependencies of the transverse Kerr effect with the DC axial magnetic field as a parameter. Two successive jumps of the magneto-optical signal in the absence of the bias field (Fig. 2.2a) reflect that two jumps of the circular magnetization take place. Under the DC axial magnetic field (H_{AXDC}) a transformation of the hysteresis loop is observed: the application of bias field causes the disappearance of one of the two jumps (Fig. 2.2b).

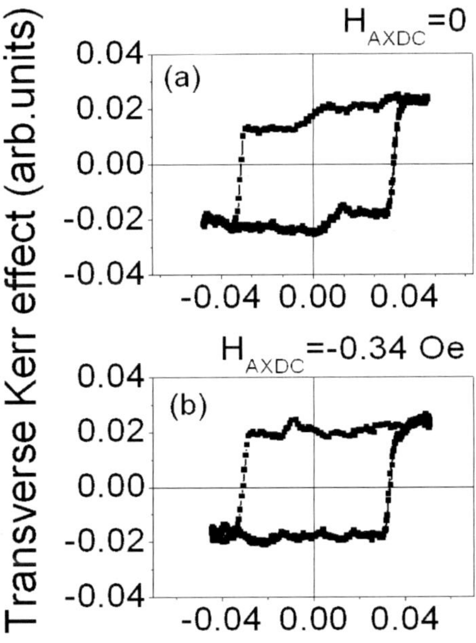

Figure 2.2 Transverse Kerr effect hysteresis loops with DC axial magnetic field as a parameter.

Considering that the Kerr effect curves contain information about the magnetization reversal in the surface area of the wire, the observed features of the magneto-optical hysteresis loop could be associated with the existence of a helical anisotropy linked to a helical domain structure in the outer shell of the wire. When the superposition of the circular and axial magnetic field is applied along the direction of this helical anisotropy, only one jump of the magneto-optical signal is observed (Fig. 2.2b). This means that only one jump occurs between two surface domains with the magnetization oriented in the helical direction for this combination of two crossed magnetic fields. The inclination of the superposed magnetic field from this direction causes the appearance of the second jump.

The value of the angle of the helical anisotropy has been estimated from the magneto-optical experiments. The value of the circular switching field (H_{SW-C}) for the first jump has been obtained from the switching current value (Fig. 2.2a) using the formula for circumferential field $H_{CIRC} = Ir/2\pi R^2$. The calculation has been performed for the surface of the wire, i.e., for $r = R$. The value of H_{SW-C} is 0.34 Oe for the value of switching current of 4 mA. Taking into account that the first jump disappears when the axial magnetic field is of 0.34 Oe (Fig. 2.2b), the value of the angle of superposition of the two magnetic fields and accordingly, of the direction of helical anisotropy, could be determined as about 45° with respect to the axial direction.

The two jumps of magnetization presented in the bulk hysteresis loop (Fig. 2.1) could also be considered the reflection of the existence of a helical anisotropy, but now in the volume of the wire. This is confirmed by the dependence of the axial switching fields on the DC circular magnetic field (DC electric current). Because the circular magnetic field has a projection on the helical direction, the jumps occur at lower axial magnetic fields in the presence of a DC circular magnetic field.

The joint analysis of the surface and bulk hysteresis loops allows us to conclude that the cold-drawing post-process induces strong helical anisotropy in the studied Fe-rich wires. The helical magnetic structure exists both on the surface and in the volume of the wire. We consider that the formation of the helical anisotropy in the studied wires is related to the deformation, which was observed on the surface of the wire after the cold-drawing process [12]. This deformation makes an angle of about 45–50° to the longitudinal direction [12].

Saturation magnetization dependence on the density of the current flowing through the wire is presented in Fig. 2.3. The insets demonstrate the change of the shape of the hysteresis loop with the heating. Two Curie points were clearly distinguishable on this dependence. This behavior of the magnetization could be associated with the formation of nanocrystalline state, which appears after the current annealing. The above-mentioned deformations on the surface of the wire, which were produced by the cold-drawing process, could be considered the driving force for the crystallization. They give the direction of the nucleation process of the nanocrystallites during current annealing. Taking into account that the jump of magnetization associated with the helical magnetic structure (Fig. 2.1) disappears during the first stage of the heating process (see insets in Fig. 2.3), we consider that the nanocrystalline phase exhibits a helical magnetic structure.

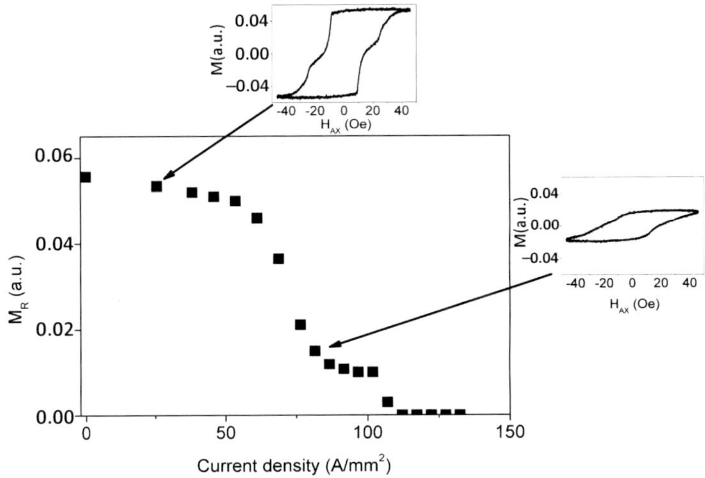

Figure 2.3 Saturation magnetization dependence on the density of the current flowing through the wire. Insets show the transformation of the hysteresis loop.

Figure 2.4 presents the AC circular magnetic field (created by an AC electric current) dependencies of the transverse Kerr effect with the DC axial magnetic field as a parameter. Figures 2.4e,f show the minor loops. Two successive jumps of the magneto-optical signal in the absence of bias field (Fig. 2.4c) reflect that two jumps of the circular magnetization take place. Under the DC axial magnetic field (H_{AXDC}) (Figs. 2.4a,b,d), a transformation of the hysteresis loop is observed. When H_{AXDC} is positive (Fig. 2.4d), the amplitude of the two jumps changes. The application of small negative bias field causes the disappearing of one of the two jumps (Fig. 2.4b). However, when the negative H_{AXDC} is high enough, the reversed second jump appears once more (Fig. 2.4a). The minor loops (Figs. 2.4e,f) demonstrate the occurrence of the first jump of surface magnetization. It is necessary to note that the minor loop obtained in the negative H_{AXDC} field is "reversed."

The direction of the second jump depends on the direction of the inclination of the superposed magnetic field from the helical direction. The transverse Kerr effect signal involves the change of the circular component of the magnetization. Therefore, the specific shape of the hysteresis loop, which is presented in the Fig. 2.4a, means that for this direction of superposed magnetic field, the direction of the remanent magnetization is closer to the axial direction than to the direction of the helical anisotropy. Therefore, the jump from this inclined remanent condition to the direction of the helical anisotropy takes place in the first stage of the magnetization reversal. This jump is shown individually in the Fig. 2.4e. The "reversed" shape of this minor loop reflects the particular increase of the circular magnetization during the magnetization reversal.

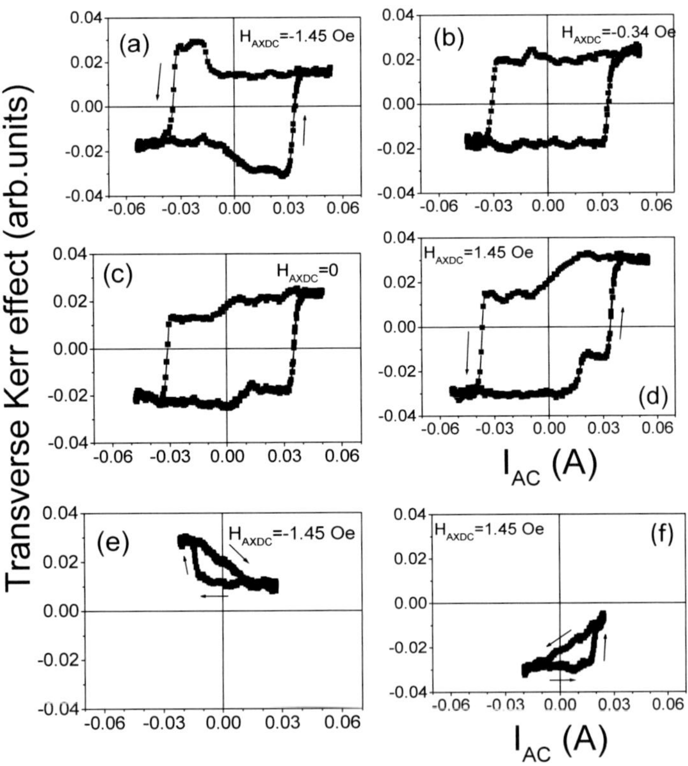

Figure 2.4 Transverse Kerr effect hysteresis loops with DC axial magnetic field as a parameter.

2.3   Helical Magnetic Structure

To study the helical magnetic structure in details, Fe-rich wires were submitted to thermal treatment by using the DC current annealing technique (220 mA for 1, 5, 15, and 45 min.). The structural characteristic of the samples was determined by the X-ray diffraction (XRD) method in a powder diffractometer provided with an automatic divergence slit and graphite monochromator using CuKα radiation ($\lambda = 1.54$ Å).

Figure 2.5 shows the influence of the DC axial bias field on the transverse Kerr effect hysteresis loop for the wire annealed for 15 min. Without the bias field, the hysteresis loop was not observed (Fig. 2.5d). Application of bias field initiates the appearance of the hysteresis loop.

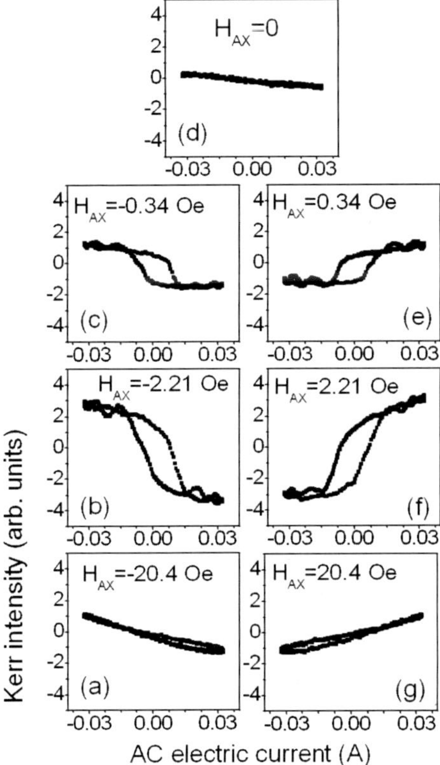

Figure 2.5   Axial bias field influence on transverse Kerr effect hysteresis loop for wire annealed 15 min.

The value of the magneto-optical signal depends on the value of the bias field. It is important to note that in the first stage, the amplitude of the magneto-optical signal (circular magnetization) increases with the axial magnetic field increase (Figs. 2.5b,c,e,f). When the bias field is high enough, the amplitude of the magneto-optical signal decreases and the hysteresis loop disappears (Figs. 2.5a,g). The shape of the hysteresis loop symmetrically depends on the sign of the bias field. The features described above have been observed also for the wires annealed for 1 and 5 min. Generally, Kerr loops can be observed at larger circular field, but it is necessary to control that the electric current does not produce essential Joule heating.

The main result obtained in this wire is the local increase of the amplitude of the circular hysteresis loops (maximal value of the circular magnetization) in the presence of the axial magnetic field (Figs. 2.5c,e,f). Usually, the DC axial field causes the decrease of the circular magnetization and the inclination of the magnetization from the circular direction toward the axial one [13]. But now we see that the DC axial field favors one of the two directions of the helical anisotropy. The observed effect means two things. First, the magnetic structure in the outer shell is a helical structure: there are two projections of the magnetization— circular and axial. Second, there is a strong correlation between the axial magnetic structure in the inner core and the helical magnetic structure in the outer shell. The reversed shape of the circular hysteresis loops (for example, Figs. 2.5b,f) reflects the correlation between the sign of axially magnetized domain, on which the DC axial field acts, and the sign of the helical domain in the outer shell (in particular, the sign of the circular projection of the magnetization in the domain).

In the absence of bias field, the transverse Kerr effect signal is almost lacking (Fig. 2.5d), which could be related to the existence of multi-domain helical structure in the outer shell when the sum projection of magnetization in the multi-domain system on the transverse axis is vanishing.

The behavior of the surface hysteresis loop for the wire annealed for 45 min has a special feature. The transformation of this hysteresis loop is asymmetric in the presence of axial bias fields of the opposite directions (Fig. 2.6): the circular saturation magnetization increases in the positive bias field (Fig. 2.6e) and decreases in the negative bias field (Fig. 2.6c). The increase in the value of the magneto-optical signal in saturation (Fig. 2.6e) reflects the rotation of the magnetization to the circular direction.

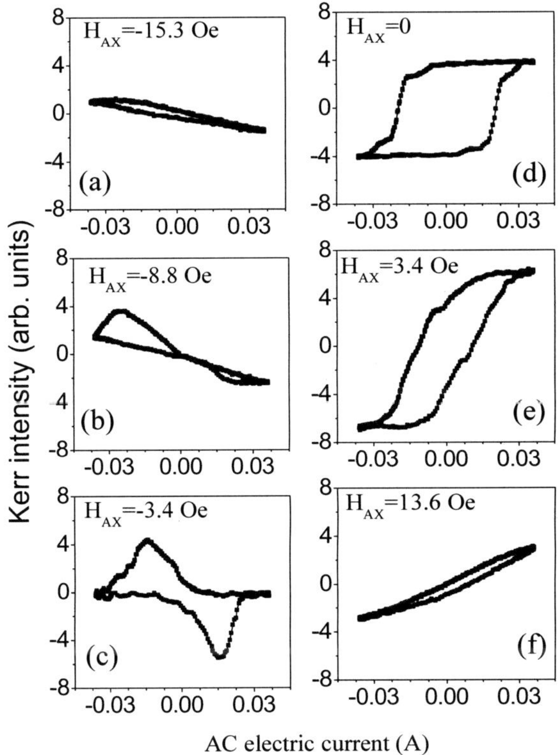

Figure 2.6 Axial bias field influence on transverse Kerr effect hysteresis loop for wire annealed 45 min.

There are two main results obtained in this wire. First, the increase of the circular saturation magnetization in the presence of the DC axial field signifies, as in the above-mentioned case, the existence of a helical magnetic structure. Second, the asymmetric transformation of the hysteresis curve means that this helical structure is mono-domain one. When the DC field is positive, the helical structure inclines toward the circular direction (Fig. 2.6e), and when the DC field is negative, the helical structure inclines toward the axial direction (Fig. 2.2c). The helical structure exists not only in the surface of the wire but also deeper. In the presence of DC axial field, we observe the inclination of helical structure as a single whole, i.e., change of the sign of the DC axial field does not cause the re-switch of the domain, as it was observed in the previous case.

Therefore, we consider the following mechanism of the helical structure formation. In the first stage of the annealing process, the multi-domain helical structure is formed in the outer shell of the wire being in the correlation with the axially magnetized inner core. The continuation of the annealing causes the formation of the mono-domain helical magnetic structure with defined direction of the rotation of the helical magnetization. This structure occupies a significant volume of the wire. We consider that the penetration of the helical structure to the volume of the wire finds the explication if we take into account the correlation of the formation of helical magnetic structure with the formation of nanocrystalline state in an amorphous matrix [8]. In this situation, the time necessary for the formation of the helical structure is determined by the time necessary for the formation of the nanocrystalline state.

The bulk hysteresis loop obtained for the wire annealed for 45 min (Fig. 2.7b) in the AC axial magnetic field (H_{AX}) demonstrates two jump of magnetization in contrast to the wire annealed 15 min (Fig. 2.7a) where the second jump is not observed. The two-jump behavior in this case is related to the existence of the helical structure in the volume of the wire [8]. The contribution of the surface to the volume hysteresis loop is small enough. Therefore, the possible jump to the helical structure in the outer shell in the wire annealed 15 min is not observed in the volume hysteresis curve.

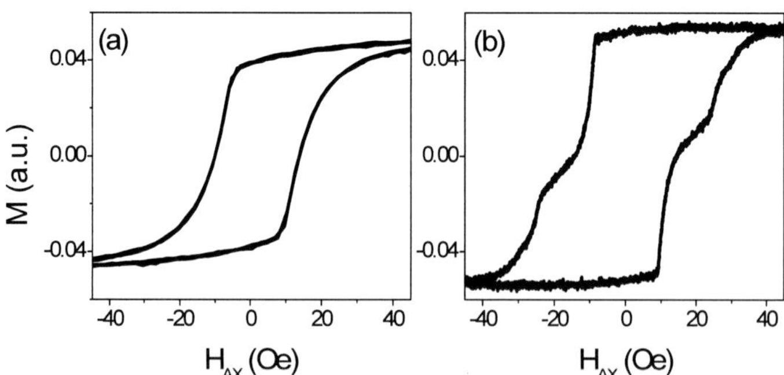

Figure 2.7 Bulk hysteresis loops for wires annealed for 15 min (a) and 45 min (b).

The results of the longitudinal Kerr effect experiment for the wire annealed for 45 min (Fig. 2.8) demonstrate that the transformation to the helical structure is accompanied by the local hysteresis that permit us to conclude that this transition is the transition of the first order. The jump to helical magnetic state is associated with the overcoming of the energy barrier related to helical anisotropy.

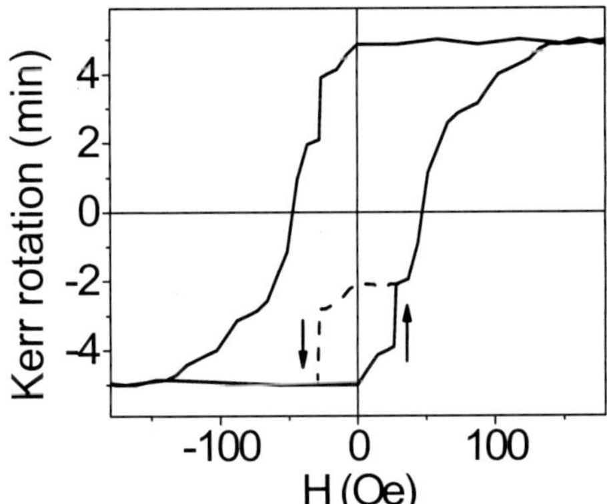

Figure 2.8   Longitudinal Kerr effect hysteresis loop for the wire annealed during 45 min. The dashed line demonstrates the local hysteresis related to the transition to helical magnetic structure.

In frame of the correlation of helical magnetic structure with the nanocrystalline state, X-ray experiments could be considered an additional confirmation of our model of the formation of helical structure. Figure 2.9 presents the raw XRD patterns for the wires

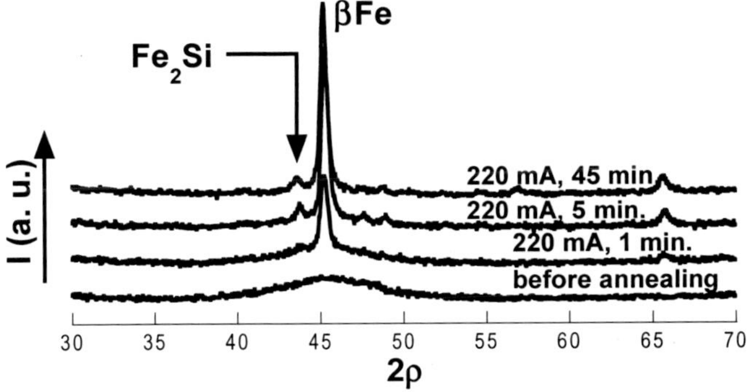

Figure 2.9   XRD raw patterns for the 50 µm diameter microwire annealed with 220 mA at different times.

submitted to current annealing for different times. The pattern for the non-annealed sample reveals the completely amorphous character of the wire. The sample crystallizes with the energy supplied by annealing: the intensity of the crystalline peaks increases with the annealing time. This indicates that the amount of the amorphous phase becomes smaller.

In the first stage of current annealing, the helical magnetic structure appears in the surface of the wire in the multi-domain form. In the second stage, the formation of mono-domain helical structure, which exists in the surface and in the volume of the wire, is observed. In addition, the observed correlation between the magnetic and the structural experiments permits us to conclude that the formation of the nanocrystalline phase is associated with the formation of the helical magnetic structure in the Fe-rich cold-drawn wire.

References

1. Malmhall R, Mohri K, Humphry FB, Manabe T, Kawamura H, Yamasaki J, and Ogasawara I (1987), *IEEE Trans. Magn.*, **23**, 3242.

2. Ogasawara I and Mohri K (1990), *IEEE Trans. Magn.*, **26**, 1795.

3. Kawashima K, Kohzawa T, Yoshida H, and Mohri K (1993), *IEEE Trans. Magn.*, **29**, 3168.

4. Freijo JJ, Vázquez M, Hernando A, Méndez A, and Remanan VR (1999), *J. Appl. Phys.*, **85**, 5450.

5. García-Beneytez JM, Vinai F, Brunetti L, García-Miquel H, and Vázquez M (2000), *Sens. Actuators A: Phys.*, **81**, 78.

6. Mohri K, Uchiyama T, Shen LP, Cai CM, and Panina LV (2002), *J. Magn. Magn. Mater.*, **249**, 351.

7. Garcia C, Gonzalez J, Chizhik A, Zhukov A, and Blanco JM (2004), *J. Appl. Phys.*, **95**, 6756.

8. Chizhik A, Garcia C, Gonzalez J, and Blanco JM (2004), *J. Magn. Magn. Mater.*, **279**, 359.

9. Chizhik A, Garcia C, Gonzalez J, and Blanco JM (2005), *J. Magn. Magn. Mater.*, **290–291**, 1472.

10. Chizhik A, Garcia C, Gawronski P, Zhukov A, Gonzalez J, Blanco JM, and Kulakowski K (2005), *J. Magn. Magn. Mater.*, **294**, e167.

11. Chizhik A, Garcia C, Gonzalez J, del Val JJ, Blanco JM, Merenkov DN, Gnatchenko SL, Shakhayeva YA, and Bludov AN (2005), *IEEE Trans. Magn.*, **41**, 3250.

12. Waseda Y, Ueno S, Hagiwara M, and Aust K (1990), *Progr. Mater. Sci.*, **34**, 149.

13. Chizhik A, Gonzalez J, Zhukov A, and Blanco JM (2002), *J. Appl. Phys.*, **91**, 537.

Chapter 3

Conventional Co-Rich Amorphous Wire

3.1 Introduction

The magnetic properties of amorphous wire prepared by an in-rotating-water quenching technique are very interesting because of their potential technological applications. Co-rich wires with nearly zero magnetostriction attract special attention, because they show a giant magnetoimpedance (GMI) effect that has been found in these wires. The importance of investigating the magnetic structures in the surface area of the wires is demonstrated by the known correlation between the GMI and the magnetic skin effect. The present chapter is devoted to the investigation of magnetic domain structure in the outer shell of an amorphous wire because of its special place in the origin of the GMI effect. The impedance is sensitive to a surface magnetic configuration at high frequencies when the skin effect is essential; therefore, the characteristic features of impedance-field behavior are closely related to a quasistatic magnetization process. During the experiments,

Magnetic Microwires: A Magneto-Optical Study
Alexander Chizhik and Julian Gonzalez
Copyright © 2014 Pan Stanford Publishing Pte. Ltd.
ISBN 978-981-4411-25-7 (Hardcover), 978-981-4411-26-4 (eBook)
www.panstanford.com

special attention was given to the behavior of the surface domains under the action of an axial magnetic field, considering that the GMI effect is very sensitive to the axial magnetic field. It was also taken into account that the transformation of the surface magnetic domain structure under the axial magnetic field has not been studied in detail in these wires.

Also, the study of the magnetic properties of Co-rich amorphous wires has been a topic of interest during the past decade because of such outstanding properties as the single and large Barkhausen jump (LBJ). The LBJ phenomenon is typically observed in highly magnetorestrictive wires obtained by the in-rotating-water method. This LBJ is caused by the magnetization reversal inside a single-domain inner cylindrical core with uniaxial magnetic anisotropy and longitudinal easy axis [1, 2]. It is supposed that the outer shell of the Co-rich wire has the magnetization with a roughly circular direction and complex multidomain structure. The existence of the circular domain structure in Co-rich wires and the possible origin of such a domain configuration has been a subject of some theoretical study [3–5]. The local easy magnetization axes are determined by the magnetoelastic anisotropy resulting from the coupling between the internal stresses and the magnetostriction. In this sense, circular magnetization measurements are quite suitable for Co-rich wires.

3.2 Magnetization Reversal in Circular and Axial Magnetic Fields

Our experiments were performed on an amorphous wire of nominal composition $Co_{72.5}Si_{12.5}B_{15}$ with a diameter of 120 μm. The transverse and longitudinal Kerr effect dependencies on the magnetic field are represented in Figs. 3.1 and 3.2, respectively. Figures 3.1a and 3.2a show the Kerr effect loops when a DC electric current does not pass through the wire. It is possible to conclude that (1) for a vanishing value of the magnetic field (in the vicinity of zero), the magnetization in the surface area of wire is oriented close to the perpendicular direction to the wire axis and (2) increasing the magnetic field amplitude is related to the rotation of the magnetization in the direction parallel to the wire axis.

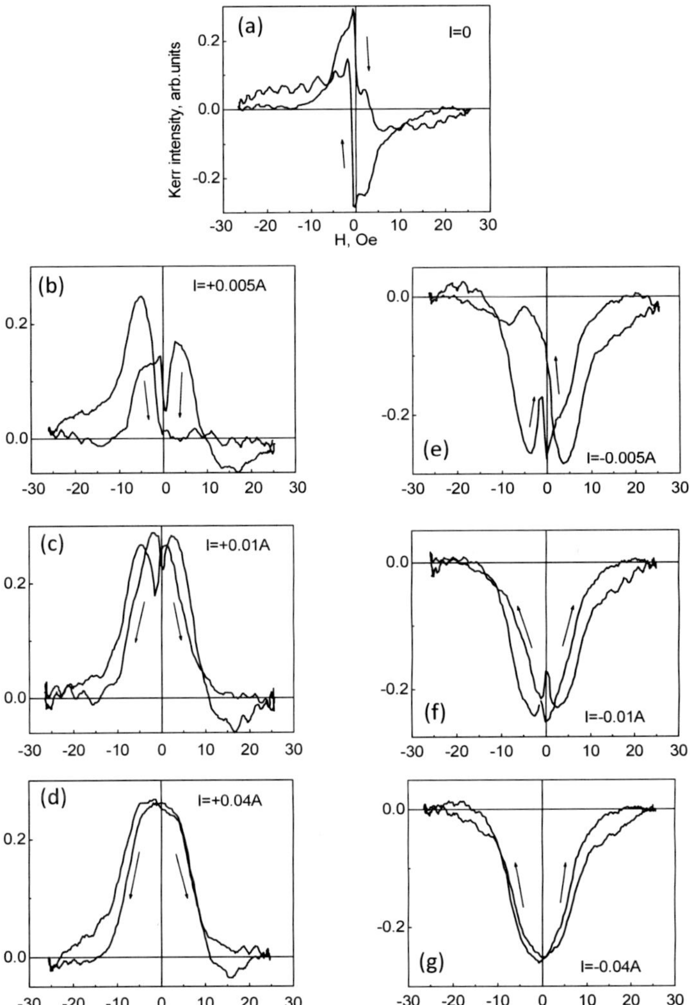

Figure 3.1 Transverse Kerr effect dependencies on the magnetic field. The values of the DC electric current flowing through the wire are (a) 0, (b) +0.005, (c) +0.01, (d) +0.04, (e) –0.005, (f) –0.01, (g) –0,04 A.

The shape of the magnetization reversal curve, presented in Fig. 3.1a has allowed us to assume the existence of circular magnetic domains with opposite magnetization directions (bamboo domains [1, 2]) in the surface area of the wire.

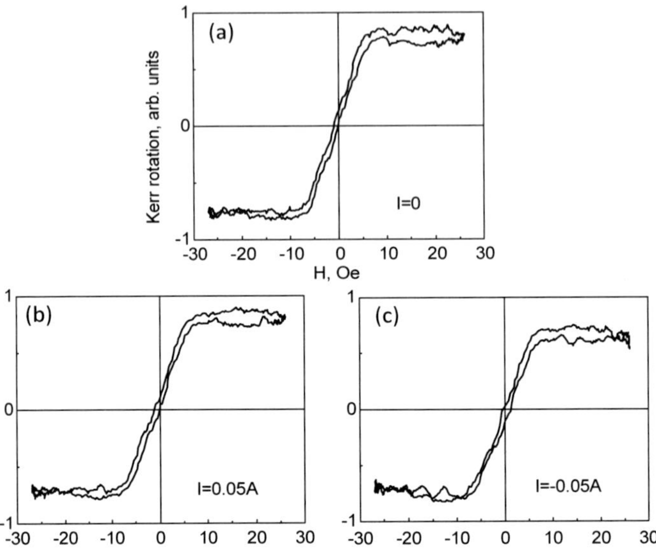

Figure 3.2   Longitudinal Kerr effect dependencies on the magnetic field. The values of the DC electric current flowing through the wire are (a) 0, (b) +0.04, (c) –0.04 A.

In addition, it was taken into account that an electric current flowing through the wire produces a circular magnetic field and consequently the circular component of the magnetization should be sensitive to the electric current [3, 4]. To check the assumption of the existence the domains in the wire studied, we have performed experiments considering the effect of a DC current flowing in the wire in addition to the magnetic field. The maximum value of the electric current has been determined by the limit above which the irreversible changes can occur in the wire because of heating. It was found that the value and the sign of the electric current strongly influenced the transverse Kerr curves (Figs. 3.1b–g). Symmetric change of the shape of the hysteresis loops has been observed depending on the sign of the current. At the same time, the influence of the current on the longitudinal Kerr effect was negligible (Fig. 3.2b,c). In addition, the Kerr effect experiments have been carried out with sweeping of the electric current instead of magnetic field sweeping.

A DC magnetic field of different signs has been applied during the experiments. The results of these experiments are presented in Fig. 3.3. When the external magnetic field is zero, the transverse

Kerr effect loop shows symmetric changes of intensity with the current change. The application of the DC magnetic field changes the shape of the loop and shifts the loop along the current axis. The sign of this shift depends on the sign of the magnetic field. The increase of the value of the magnetic field causes a decrease of the value of the transverse Kerr intensity until the loop disappears.

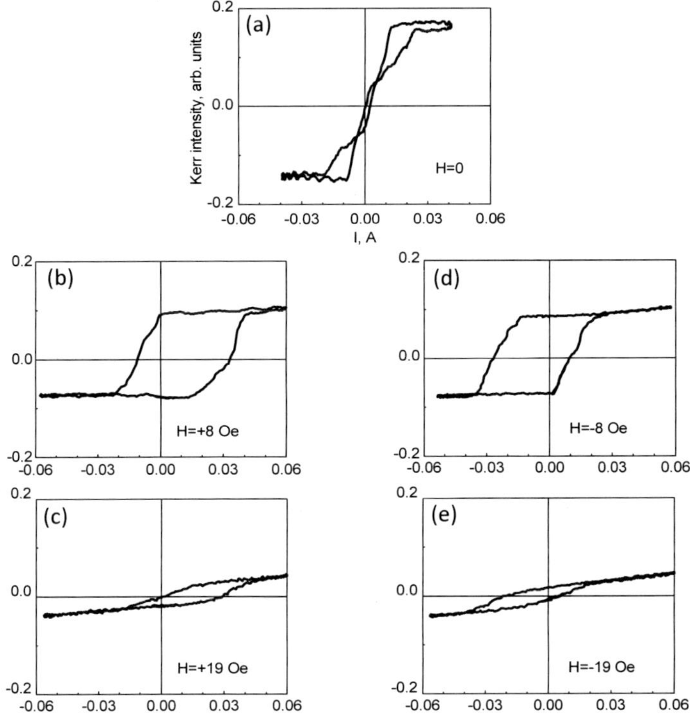

Figure 3.3 Transverse Kerr effect dependencies on the electric current flowing through the wire. The values of the DC magnetic field are (a) 0, (b) +8, (c) +19, (d) –8, (e) –19 Oe.

3.3 Model of Surface Magnetization Reversal

The analysis of the above experiments can be performed using a model, which takes into account the existence of circular magnetic domains in the outer shell of the wire. It is based on the simultaneous change of the relation of the volumes V_1 and V_2 of the domains with different direction of the circular magnetization and the rotation of the magnetization in the domains (Fig. 3.4). The equal number of the domains with different direction of the magnetization under the light spot was assumed. A similar model has been successfully applied to the magnetization reversal in metallic multilayers [6].

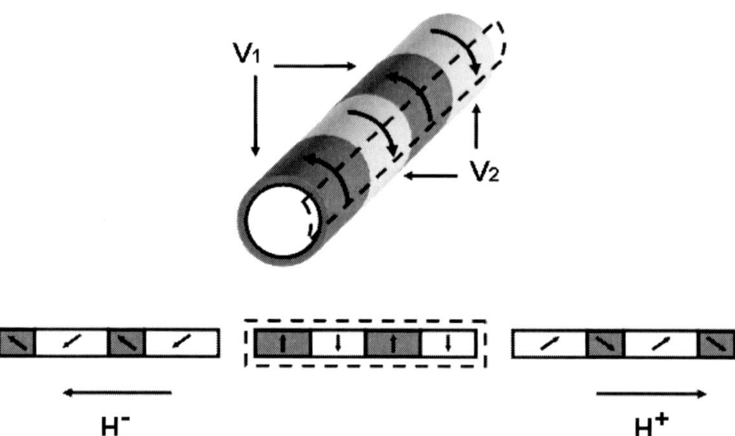

Figure 3.4 Model of circular magnetic domains in the outer shell of the wire. The dotted line schematically shows the area of the wire surface, which reflects the light. The lower pictures show schematically the simultaneous change of the relation of volumes V_1 and V_2 of the domains and the rotation of the magnetization inside domains.

As is known, the transverse Kerr effect reflects the change of the part of the magnetization, which is perpendicular to the field-light plane (perpendicular magnetization). We consider the field dependencies of the perpendicular component of the magnetization $M_{1\text{perp}}(H)$ and $M_{2\text{perp}}(H)$ in order to compare calculation and experiment.

Modeling our experimental results, we consider the field dependencies of the perpendicular magnetization $M_{1\text{perp}}(H)$ and $M_{2\text{perp}}(H)$ inside volumes V_1 and V_2 taking into account the field dependencies of the volumes $V_1(H)$ and $V_2(H)$. Accordingly, the following expression can be written:

$$M_{\text{perp}}(H) = M_{1\text{perp}}(H) \times V_1(H) + M_{2\text{perp}}(H) \times V_2(H) \tag{3.1}$$

$M_{\text{perp}}(H)$ is the field dependence of the perpendicular magnetization in the system with the change in the relation of the domain volumes and the rotation of the magnetization in the domains.

Based on the analysis of the DC electric current influence to the magnetization reversal (Fig. 3.1), the shape of the $M_{1\text{perp}}(H)$ and $M_{2\text{perp}}(H)$ dependencies has been chosen for modeling. The current influence produces additional change in the relation between the volumes V_1 and V_2. When the current is high enough (more than 0.04 A), there is only one circular domain. In fact, the transverse Kerr effect loops, which are shown in Figs. 3.1d (I = +0.04 A) and 3.1g (I = –0.04 A) reflect the field dependencies of the perpendicular magnetization in one of the domain (in V_1 or V_2 depending on the sign of the current). Therefore, these two field dependencies of the Kerr effect can be taken as the $M_{1\text{perp}}(H)$ and $M_{2\text{perp}}(H)$ dependencies in the model.

The results of the first stage of the modeling are presented in Fig. 3.5. The field dependencies of the domain volumes $V_1(H)$ and $V_2(H)$ were examined to be in agreement with the experimental transverse Kerr and calculated $M_{\text{perp}}(H)$ loops. It was assumed that V_1 changed from 0 to V_{sum} and V_2 changed from V_{sum} to 0 under the condition $V_1 + V_2 = V_{\text{sum}}$ = Const. As a result, the calculated $M_{\text{perp}}(H)$ dependence (Fig. 3.5e) reflects the main features that are observed on the experimental transverse Kerr curve (Fig. 3.1a). The smooth increase and decrease of M_{perp} could be associated with the rotation of the magnetization inside V_1 or V_2. The sharp change of the magnetization in the vicinity of the zero field is related to the sharp change in the relation between volumes V_1 and V_2. Different signs of M_{perp} in the different parts of the curve mean that the magnetization value in V_1 is larger than the magnetization value in V_2 or vice versa depending on the domain volume relation. The effect of hysteresis, observed in Fig. 3.5e is related to the proposed hysteresis of the domain volume change (Figs. 3.5a,b).

In the second stage of our modeling, we took into consideration the influence of the DC electric current to the domain volume relation. Generally, V_1 volume changes from $K_1 \times V_{\text{sum}}$ to $K_2 \times V_{\text{sum}}$ and consequently V_2 volume changes from $(1 - K_1) \times V_{\text{sum}}$ to $(1 - K_2) \times V_{\text{sum}}$. The values of the K_1 and K_2 coefficients depend on the value and sign of the current. The method of the $M_{\text{perp}}(H)$ calculation was the same as in the first stage of the modeling. The results of the calculations are presented in Fig. 3.6 in the form of a comparison with the experiment.

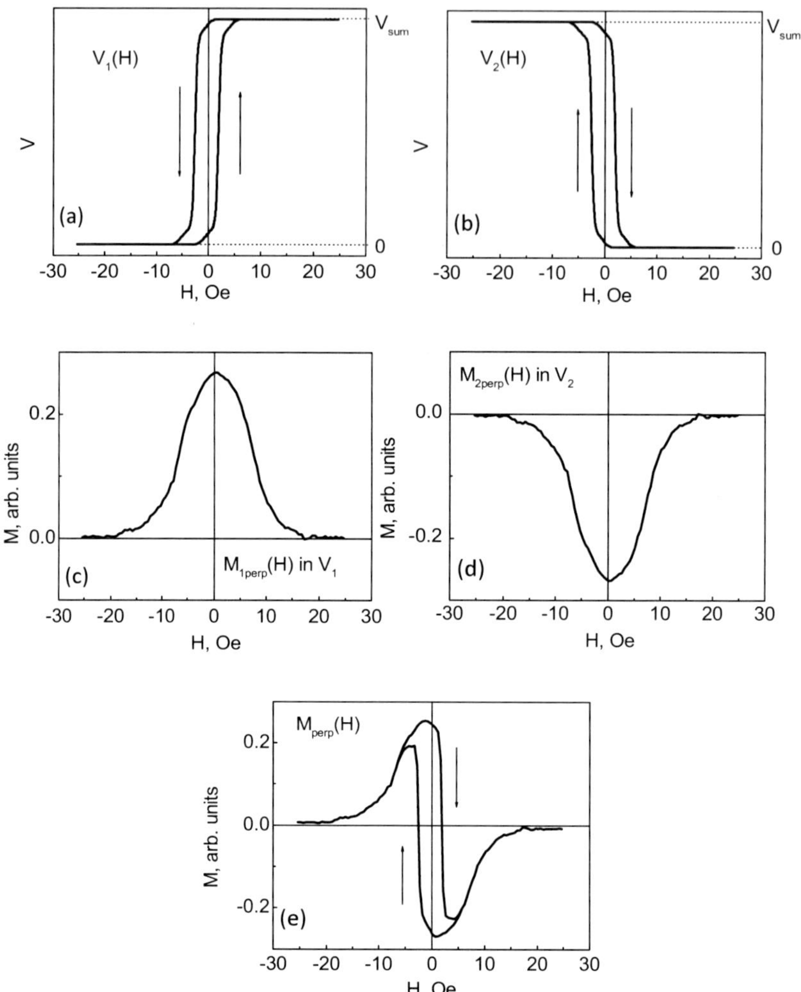

Figure 3.5 Modeling of the magnetization reversal in the outer shell of the wire when a DC electric current does not flow through the wire. Figures (a) and (b) present the field dependencies of the domain volumes $V_1(H)$ and $V_2(H)$; (c) and (d) present the field dependencies of perpendicular magnetization $M_{1perp}(H)$ and $M_{2perp}(H)$ inside volumes V_1 and V_2; (e) presents the calculated $M_{perp}(H)$ dependence.

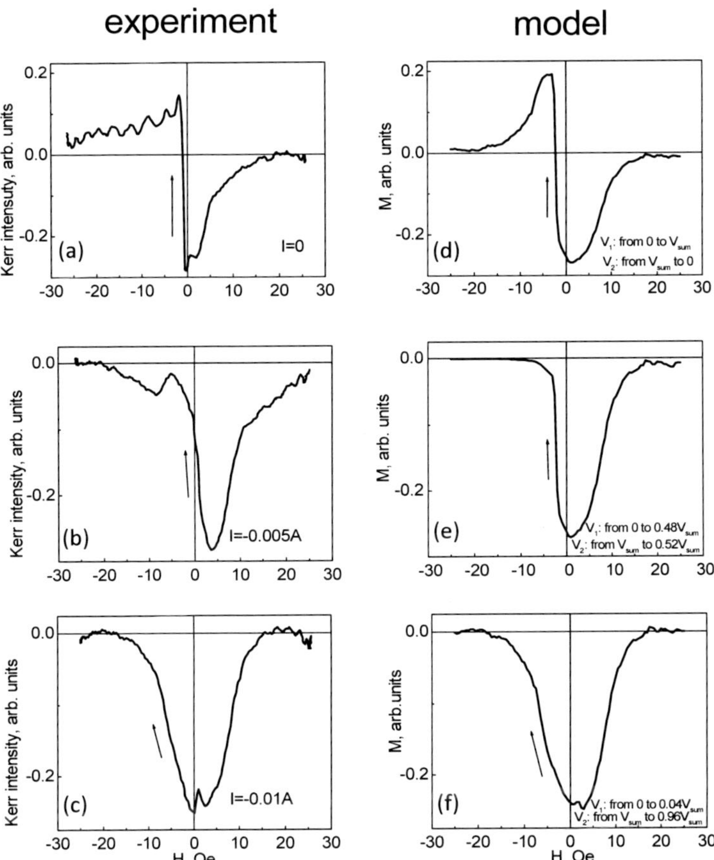

Figure 3.6 Comparison of the experiment and the calculation when a DC electric current flows through the wire. Experiment: (a) $I = 0$; (b) $I = -0.005$; (c) $I = -0.01$ A. Calculation: (d) $K_1 = 0$, $K_2 = 1$; (e) $K_1 = 0$, $K_2 = 0.48$; (f) $K_1 = 0$, $K_2 = 0.04$.

Figures 3.6a–c present the half-loops of the experimental transverse Kerr effect, which were taken from Fig. 3.1. Figures 3.6d–f present the calculated curves for different values of the K_1 and K_2 coefficients. The values of K_1 and K_2 were chosen to obtain the best agreement between experiment and calculation. From the comparison of the experiment and the calculation, it was possible to conclude that the amplitude of the change of one domain increases when such amplitude of the change of another domain decreases in the case when an electric current is flowing in the wire. Therefore, the signal from one domain should increase and the signal from another domain should decrease when the DC current increases. Finally, when one of the domains disappears, the Kerr effect dependencies have shapes as shown in Figs. 3.1d,g. It is interesting to note that a little fall of the signal (which is obtained both in the experiment and in the modeling (Figs. 3.6c,f) can be explained as a contribution coming from the small, disappeared domain. In fact, when one of the domains disappears under the action of the electric current, the fall disappears, too.

The shape of the longitudinal Kerr effect loops (Fig. 3.2) can also be explained in the framework of the proposed model. The projection of the magnetization parallel to the field-light plane has the same sign in the domains V_1 and V_2 (Fig. 3.4). The longitudinal Kerr effect, which is proportional to the parallel projection of the magnetization, is, therefore, not sensitive to the change of the relation between V_1 and V_2 volumes.

Moreover, we have examined the transformation of the current dependencies of the Kerr effect (Fig. 3.3) from the point of view of possible existence of bamboo domains in the wire surface. For the external DC magnetic field equal to 0 (Fig. 3.3a), the Kerr loop reflects the changes of the domain volume relation without rotation of the magnetization. The area on the Kerr curve for the current of more than 0.02 A and less than −0.02 A, corresponds to the existence of only one domain. The application of DC magnetic field in addition to the AC current changes the shape of the loop and shifts the loop. The effect of the DC magnetic field is associated with changes in the nucleation mechanism of the domains. The main cause of the shift of the Kerr loop is that the DC magnetic field assists nucleation of the circular domains of one type and inhibits nucleation of the domains of the other type. This correlation of the shift direction and sign of the magnetic field is strong evidence confirming our explanation. The following decrease of the Kerr intensity with the disappearance of the loop, which takes place as the magnetic field increases, can be attributed to the progressive inclination of the magnetization toward direction of the magnetic field, decreasing the perpendicular projection M_{perp} of the magnetization.

To summarize, we have demonstrated that the Kerr effect is a useful tool to investigate the magnetization reversal in amorphous magnetic wires. The experiments performed and their analysis have shown that the magnetization reversal in the surface area of a Co-rich amorphous wire is complex owing to the complicated magnetization process. The use of Kerr effect in the presence of both magnetic field and electric current allowed us to elucidate the main features of this process. The good agreement between the experiment and the model is worth mentioning . This model is based on the formation of bamboo domains, changes of the relation of the domain volumes, and the rotation of the magnetization in the domains.

3.4   Effect of a Thermal Treatment

The effect of thermal treatment has been studied in amorphous wire of nominal composition $(Co_{94}Fe_6)_{72.5}Si_{12.5}B_{15}$ (the diameter is 120 μm) obtained by the in-rotating-water quenching technique. This amorphous wire was submitted to thermal treatment without torsion and under torsion $((\pi/4)$ rad/cm), by using the current annealing technique at 450 mA (current density 36 A/mm^2) for 20 min. During the current annealing, the wire was heated above the Curie point that was controlled by the disappearing of the hysteresis loop. The experiments have been performed in as-quenched and annealed (without and under torsion) samples.

3.5   As-Quenched Wire

There were four variants of experiments, depending on the combination of the magnetic field and type of the Kerr effect:

- transverse Kerr effect as a function of circular field (± axial bias field);
- longitudinal Kerr effect as a function of circular field (± axial bias field);
- transverse Kerr effect as a function of axial field (± circular bias field);

- longitudinal Kerr effect as a function of axial field (± circular bias field).

Figure 3.7 presents the transverse (a–d) and the longitudinal (e–h) Kerr effect dependencies on the circular magnetic field produced by the current, I, flowing along the sample, at the presence of DC axial magnetic field (H_{AX}). The values of DC axial field are marked in the figures. When $H_{AX} = 0$, the transverse hysteresis loop is enough rectangular with the sharp vertical areas related to the magnetization reversal process [7]. Significant modifications of the hysteresis loop have been observed under application of axial field: in the presence of axial field two jumps of magnetization take place. The longitudinal hysteresis loop exhibits two sharp peaks in the absence of axial field, which also are modified under the action of the axial magnetic field. The comparison of transverse and longitudinal loops permits us to clarify the mechanism of magnetization reversal of the wire. We assume a formation of circular bamboo domains during magnetization reversal for the studied as-quenched wire. Changes of the shape of the magneto-optical signal on the loops (b) and (f) can be ascribed to the quick rotation of magnetization and nucleation of the new circular domains. In the presence of axial field there is only one circular domain and from Figs. 3.7e,g,h the rotation of magnetization can be assumed. The direction of the rotation depends on the sign of the DC axial field. Such rotation consists of two jumps.

Hysteresis loops obtained under AC axial field provide an additional confirmation of this model of circular multi-domain system (Fig. 3.8). The values of DC current flowing through the wire and producing DC circular magnetic field are marked in the figure. In absence of the bias circular field, the transverse Kerr effect signal is almost lacking because the sum projection of magnetization in the circular multi-domain system on the transverse axis during magnetization rotation is vanishing. When the DC circular field is applied, the relation between volumes of circular domains is changed and, therefore, the sum projection increases. The sign of the projection depends on the sign of circular field.

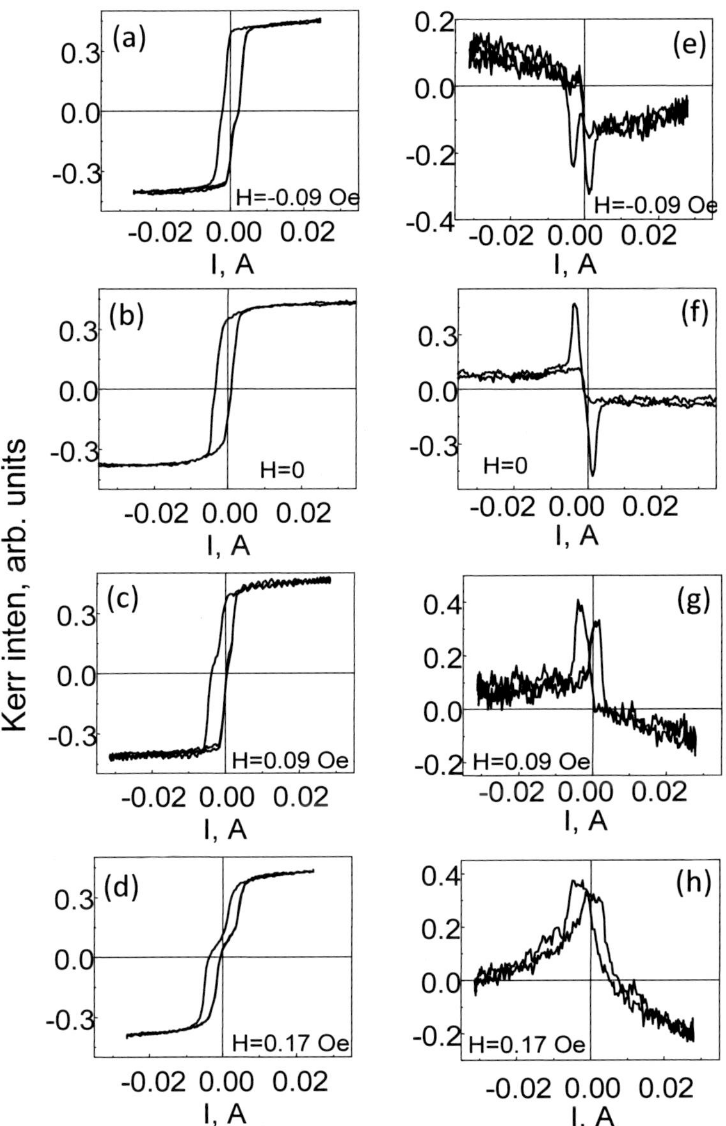

Figure 3.7 Kerr effect hysteresis loops of as-quenched Co-rich wires, which were obtained in AC circular magnetic field in presence of DC axial magnetic field. (a–d): transverse Kerr effect; (e–h): longitudinal Kerr effect.

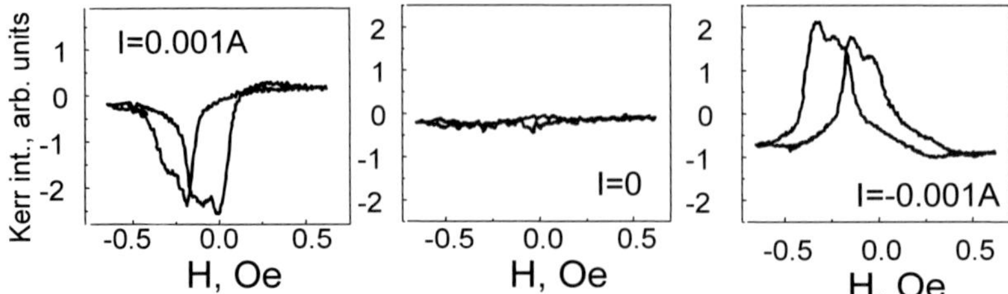

Figure 3.8 Transverse Kerr effect hysteresis loops of as-quenched Co-rich wire which were obtained in AC axial magnetic field in presence of DC circular magnetic field.

3.6 Annealed Wire

Figure 3.9 shows the longitudinal and transverse Kerr effect dependencies on the circular magnetic field with the DC axial magnetic field. The values of DC axial field are marked in the figures. Significant changes of hysteresis loops have been observed, as compared with as-cast sample. The shapes of the observed curves have been examined using the model, which takes into account formation of a maze domain structure with closure domains in the outer shell, during magnetization reversal. It is known that the magnetization in the closure domains is aligned parallel to the axis of the wire. The magnetization reversal occurs as a successive magnetization rotation and domain nucleation [6]. The monotonic increase of the signal with increase of the current from −0.2 A (longitudinal curve in Fig. 3.9a) is related to the rotation of magnetization. We assume that rather sharp changes of magnetization are associated with nucleation of new domains, when the magnetization in the outer shell is parallel to the axis of wire. The nucleation of domains has not been observed in the transverse curves because the sum of the transverse projections of the magnetization inside the domain during magnetization reversal does not depend on the relation of domains volumes. The evolution of hysteresis, presented in Figs. 3.9b–d shows how under the action of DC axial field the equilibrium relation between domains changes up to the existence of only one domain. In this case, only the rotation of magnetization is observed.

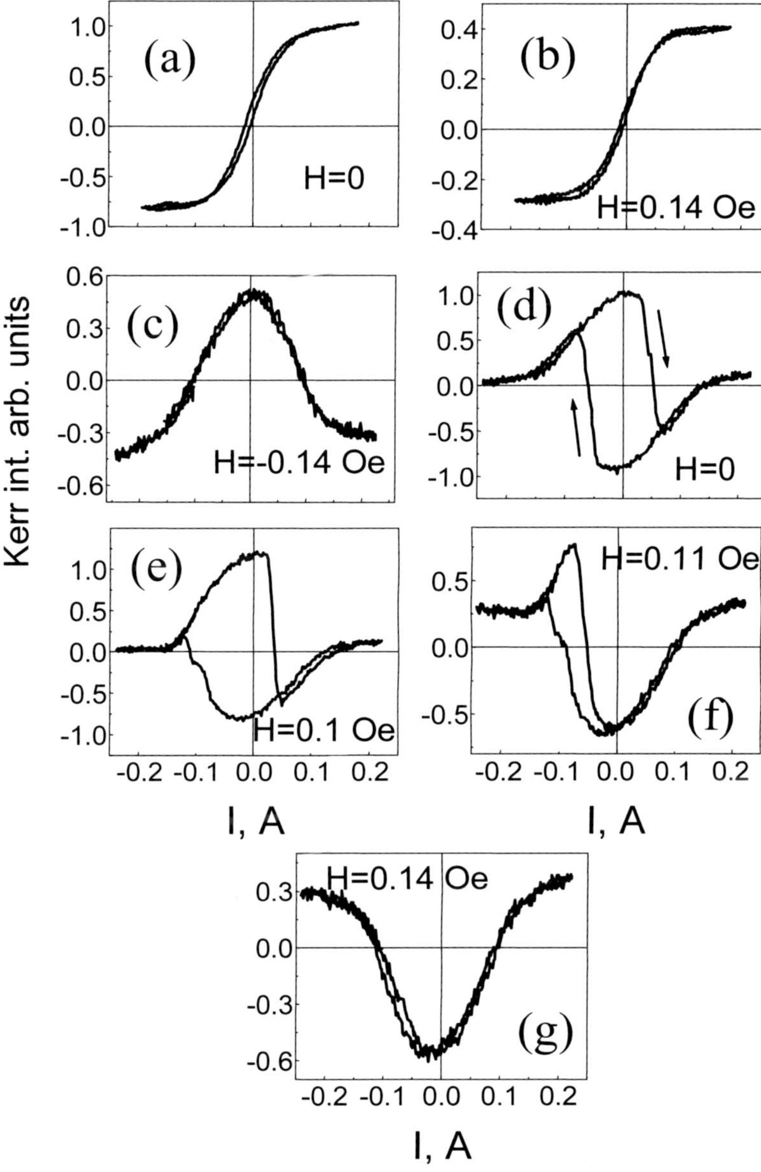

Figure 3.9 Longitudinal and transverse Kerr effect hysteresis loops of annealed Co-rich wire which were obtained in AC circular magnetic field in presence of DC axial magnetic field.

3.7   Torsion-Annealed Wire

Longitudinal and transverse Kerr effect loops obtained in axial and circular field are presented in Fig. 3.10 for the torsion-annealed wire. All the four kinds of the experiments show that jumps related to the quick change of magnetization are presented. We suppose that torsion annealing develops helical structure in the wire. For this configuration, the magnetization within the wire has two projections: axial and circular. Therefore, the jump is observed between two states: (1) axial and circular magnetization "+" and (2) axial and circular magnetization "–". The jump is followed by the successive rotation of the magnetization to the field (axial or circular) direction. Obviously, the character of magnetization reversal of the wire with helical magnetic structure in circular and axial field results to be quite similar because the magnetization in this case is sensitive equally to both axial and circular magnetic field.

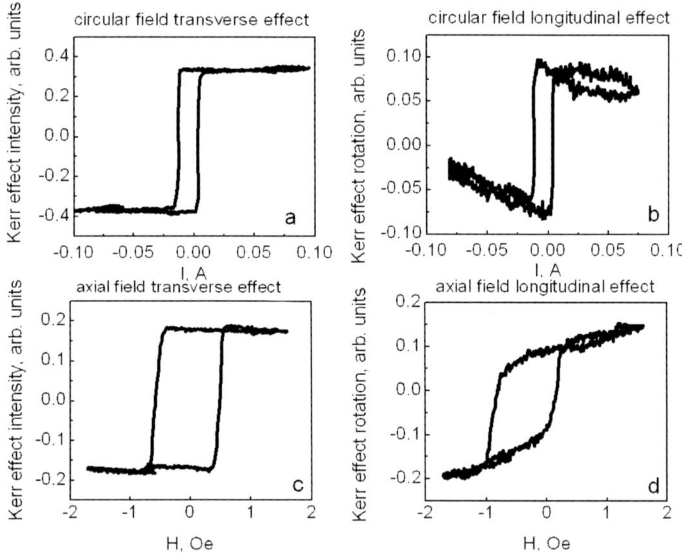

Figure 3.10   Longitudinal and transverse Kerr effect loops of torsion annealed Co-rich wires that were obtained in AC axial and AC circular field.

The magneto-optical investigation shows that the drastic transformation takes place in the domain structure of amorphous wires under the annealing. The observed changes of Kerr effect loops could be related with the change of the domain structure from circular type (as-quenched sample) to longitudinal type (annealed sample) on the wire surface. According to Refs. [7, 8], we attribute this modification to the change of the magnetostriction constant sign.

To confirm our assumption, we have evaluated the saturation magnetostriction constant (λ) by the small-angle magnetization rotation method obtaining the following values of λ: for as-quenched wire $\lambda = -4.2 \times 10^{-8}$ and for annealed wire $\lambda = 3.2 \times 10^{-7}$. As it was assumed, the sign of the magnetostriction constant has been changed after annealing. This behavior has been widely explained by the fact that current annealing (without torsion stress) gives rise to a relaxation of the complex internal stresses induced during the fabrication process and the decrease of the magnetoelastic anisotropy associated with such frozen-in stresses. In these conditions, the magnetostrictive properties in the annealed wire are largely determined by the chemical composition of the wire.

Torsion annealing induces a helical magnetic anisotropy that is displayed in our Kerr effect results. The quick changes of the magnetization observed in AC circular and AC axial magnetic fields are similar to the large Barkhausen jump inside the helical magnetic structure.

3.8 Correlation between Switching Field and Wire Length

The influence of the wire length has been studied in the wire of composition $Co_{72.5}Si_{12.5}B_{15}$ with a diameter of 120 mm. Samples of wire with length of 2.5, 5, and 10 cm have been studied. The transverse Kerr effect dependencies on the electric current flowing along the wire are presented in Fig. 3.11. Figure 3.12 compares the one-way half-loops for the set of wires with different length. We have analyzed the obtained results assuming the existence of circular magnetic domains with opposite magnetization direction. Nucleation, annihilation of the domains, and propagation of domain walls in the outer shell of wire have been assumed. The following features of the Kerr dependencies should be mentioned. The saturation areas for large values of electric current observed in the loops are associated with the existence of only one circular domain. The area of the loops related to the magnetization reversal consists of two parts with a different slope. The relation of these two parts changes with the wire length. Besides this fact, the value of the switching current (and accordingly, the value of the circular switching field) increases as the wire length increases. In the performed experiments, we assume that the switching field should be the field in which the nucleation of circular domain starts [9].

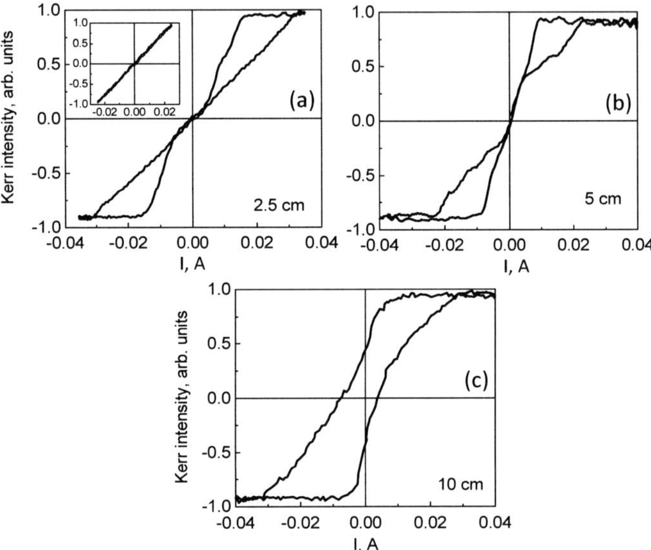

Figure 3.11   Transverse Kerr effect dependencies on the electric current flowing through the wire for different wire length: (a) 2.5, (b) 5 and (c) 10 cm.

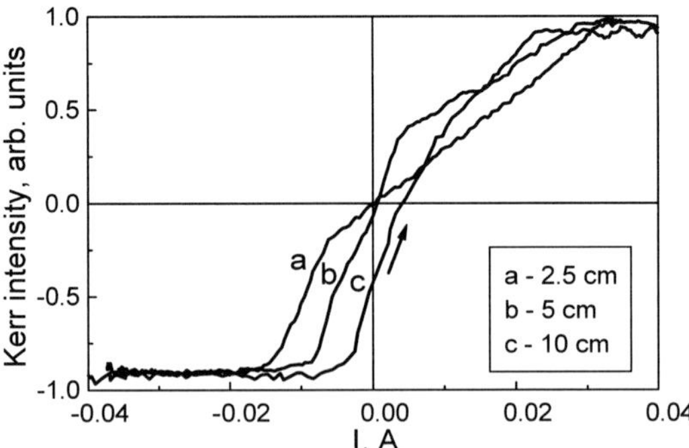

Figure 3.12   Comparison of the one-way half-loops for the set of wires of different length.

Taking into account that the circular domain width depends on the relation between the length and diameter of the wire, the increase of the switching current could be explained owing to the increase of the domain width with the wire length. The growth of the critical size of a nucleated domain requires an increase of the switching magnetic field that is reflected in the shift of the switching current. The first part of magnetization reversal, relatively sharp, is associated with a nucleation process, while the second part is related to a slower process of domain wall propagation.

The insert of Fig. 3.11 shows a smooth minor loop reflecting a regular motion of the domain walls. Increase of the electric current causes the appearance of hysteresis related to the nucleation of domains. Therefore, the magnetization reversal starts from the overcoming of the energetic barrier associated with the nucleation. Because the value of the barrier is different in samples with different lengths, the value of switching current changes with the wire length. The second step of the remagnetization—regular domain wall propagation—starts when the domain width reaches the equilibrium volume corresponding to a given external magnetic field and the nonequilibrium process of nucleation is replaced by a equilibrium process of domain wall motion. Besides this, we have performed experiments when an axial DC magnetic field has been additionally applied to an AC circular magnetic field created by sweeping electric current.

Successive influence of the DC magnetic field on the Kerr effect loops is presented in Fig. 3.13 and summarized in Fig. 3.14. It can be observed that the part of the loop, related to the nucleation of domains, increases with the DC magnetic field. The angle of slope grows and the shape of Kerr loop changes to be a quite rectangular loop. This rectangular shape of the loop is related with the character of the magnetization reversal similar to LBJs. A weak shift of Kerr loops, presented for the wire with a length of 2.5 cm in Fig. 3.13, has been observed more expressively for the length of 5 cm. The shift direction correlates with the sign of the magnetic field. The main cause of the transformation and shift of the loops is that the DC magnetic field favors the nucleation of a circular domain of one type and delays the nucleation of the domains of the other type. The influence of an axial magnetic field on the nucleation of circular magnetic domains is determined by the correlation between the circular and axial components of the magnetization of the wire. Such correlation has been proposed as the reason of a similar shift observed in stress annealed Co-rich wire. This effect of the circular hysteresis loop shift under the action of an axial magnetic field has found confirmation in our Kerr effect experiments for a more common case of as-cast wires. It is also worth mentioning that the switching field dependence on the length of the wire is associated with the correlation between the axially magnetized inner core and circularly magnetized outer shell. In this way, the sample length affects the circular magnetization process through the change of the axial demagnetizing field.

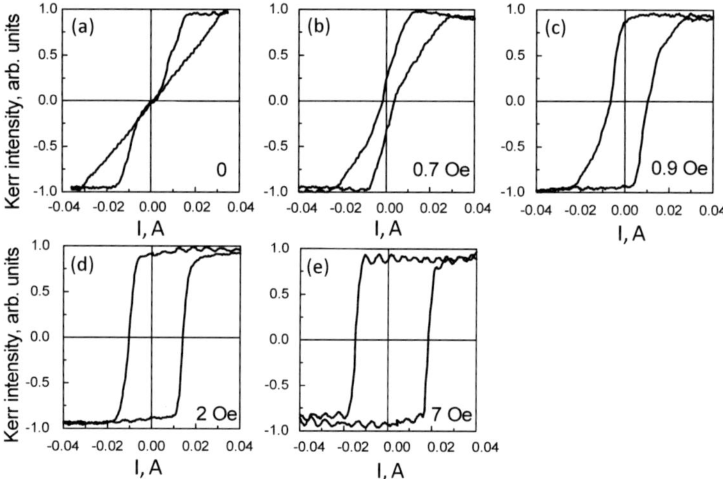

Figure 3.13 Transverse Kerr effect dependencies on the electric current for different value of DC magnetic field: (a) 0, (b) 0.7, (c) 0.9, (d) 2 and (e) 7 Oe. The length of wire is 2.5 cm.

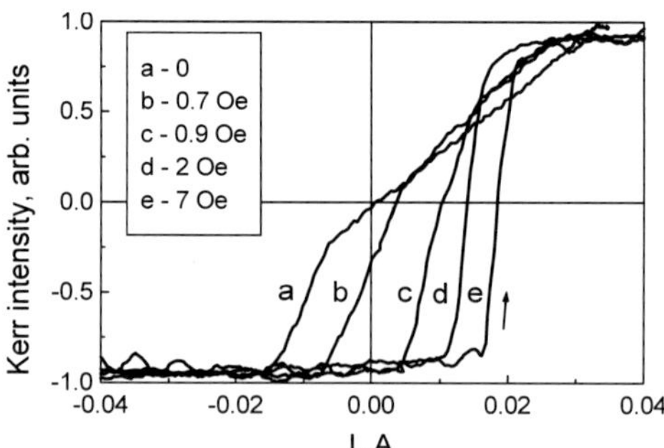

Figure 3.14 Comparison of the one-way half-loops for different values of a DC magnetic field. The length of wire is 2.5 cm.

A following remarkable transformation of the loops with the increase of absolute value of the DC magnetic field was observed. The amplitude of the Kerr loop decreases with a further disappearing of the loop. This behavior could be originated by the progressive inclination of the circular magnetization component toward the direction of the axial magnetic field.

Comparing the influence of the DC axial magnetic field on the transverse Kerr effect loops for 2.5 and 5 cm, it is possible to make some suggestions related to the observed transformation of the domain structure. In the first stage of the transformation (magnetic field 8 Oe, see Fig. 3.13), the DC axial field destroys the multidomain structure and creates a single domain circular state. It is supported by the observation of the rectangular Kerr effect loop without decreasing of the absolute value of the Kerr signal (Fig. 3.13e). In the second stage (axial magnetic field, 8 Oe < H < 19 Oe), the absolute value of the Kerr signal decreases maintaining the rectangular shape of the hysteresis loop. This means that there are magnetization jumps between two states of the outer shell with the magnetization slightly inclined from the circular direction toward the axial direction. In the third stage (axial magnetic field above 19 Oe), the fluent rotation between two strongly inclined magnetic states is observed. Concluding, Kerr effect investigations have been performed in a Co-rich amorphous wire in the presence of a circular magnetic field created by an electric current flowing along the wire. It was found that the magnetization reversal process in the outer shell consists of two parts associated to circular domain nucleation and domain wall propagation. A strong influence from a DC axial magnetic field on the Kerr effect loops has been observed. The influence is explained considering the correlation between the inner core and the outer shell of the wire.

3.9 Vortex-Type Domain Structure

The investigations of vortex domain structure have been performed in wires of nominal composition $(Co_{94}Fe_6)_{72.5}Si_{12.5}B_{15}$ (diameter 120 μm). The length of the studied wires was 7 cm. The process of magnetization reversal in the surface area of the wires has been studied by a magneto-optical Kerr effect loop tracer and by a Kerr microscope employing an image processor.

Figure 3.15 presents the transverse Kerr effect dependence on the electric current, *I*, flowing through the wire and producing the circular magnetic field. In addition, Fig. 3.15 presents the domain patterns obtained by the Kerr microscope. The magnetization reversal appears between two states with opposite directions of circular magnetization in the outer shell of the wire. The "black" and "white" colors correspond to these two opposite directions of the circular magnetization. It is possible to observe the nucleation of circular domains with successive domain wall propagation (Fig. 3.15b) and the formation of the domain structure of bamboo type (Fig. 3.15c). The change of the domain walls shape takes place during the domain wall propagation. This is related to pining effect of the domain walls.

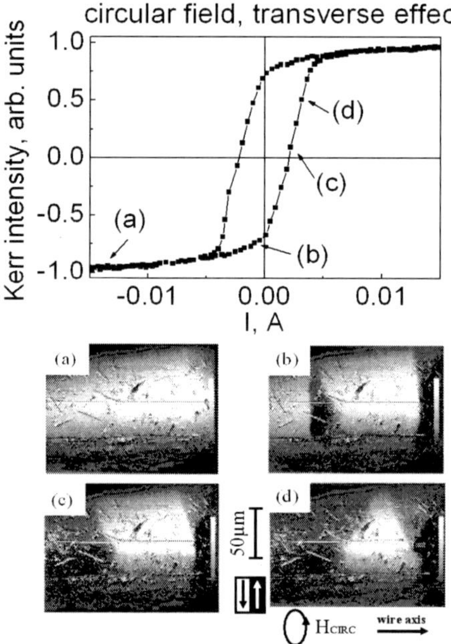

Figure 3.15   Transverse Kerr effect hysteresis loop obtained in circular magnetic field and images of surface domain structure.

If the behavior of surface magnetic structure in the circular magnetic field looks as is predicted taking into account the circular anisotropy in the outer shell, the results of magneto-optical experiments for the case of axial magnetic field are unexpected. Figure 3.16 presents the hysteresis loop and the domain patterns obtained when the external magnetic field was applied along the wire axis, that is, perpendicularly to the direction of circular surface anisotropy. In the first stage of the magnetization reversal process (Fig. 3.16b), the change of magneto-optical contrast is observed in some areas that can be attributed to the rotation of the magnetization. Further, the complex multi-domain structure appears (Fig. 3.16c). The transformation of this domain structure is accompanied by the domain walls motion (Figs. 3.16d,e). In the last stage of the magnetization reversal process, fluent change of the contrast also takes place (Fig. 3.16f).

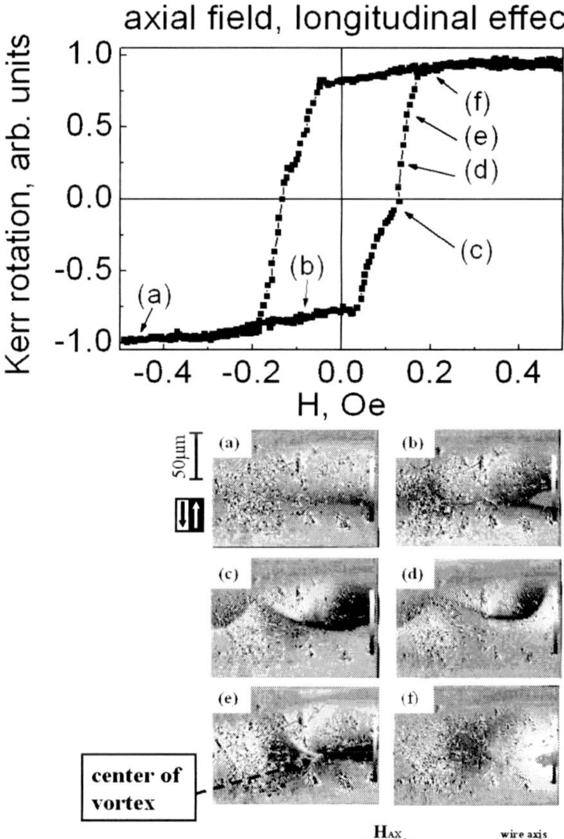

Figure 3.16 Longitudinal Kerr effect hysteresis loop obtained in axial magnetic field and images of surface domain structure.

Analyzing the results obtained, we used the schematic pictures of domain structure in the outer shell of the wire (Fig. 3.17).

When the absolute value of the axial magnetic field is high, the magnetization is directed along the wire axis (Figs. 3.17a,f). These conditions are depicted as "gray" (Fig. 3.16a). When the magnetic field decreases, the change of the contrast happens in the wire surface and the formation of some modulated structure is observed (Fig. 3.16b). The axial projection of the magnetization in these pre-domain states is equal, but the circular projections differ from each other (Fig. 3.17b). The appearance of domains with the opposite axial direction of the magnetization (Fig. 3.16c) is the second stage of the magnetization reversal. This moment is shown schematically in Fig. 3.17c. This appearance is reflected in the hysteresis loop (Fig. 3.16) as a sharp change of the magneto-optical signal. Thus, domains of four types exist in this stage. They are marked as 1, 2, 3, and 4. The domain walls between domains 1, 3 and 2, 4 are clearly observed, but the positions of domain walls between domains 1, 2 and 3, 4 are not so evident. It should be related to the value of the angle Φ, at which the magnetization rotates in the domain walls. There are two types of domain walls. The angle Φ of the domain walls of the first type is near 180° and it is easily observed. The angle Φ of the domain walls of the second type is small and determined by the inclination of the magnetization from the axial direction. During the magnetization reversal, fluent domain wall motion and jump-like rearrangement of whole domain structure takes place. In this way, domains 3, 4 replace domains 1, 2 (Figs. 3.16e,f). Further, the magnetization in domains 3, 4 rotates towards the axial direction and the contrast disappears.

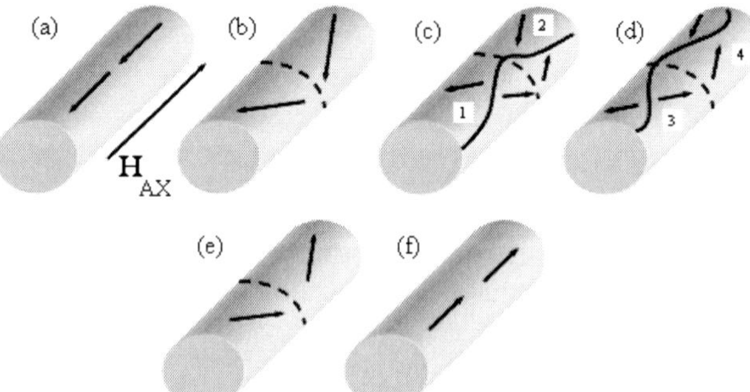

Figure 3.17 Schematic diagram of the evolution of the surface domain structure in the axial magnetic field. Arrows show directions of the magnetization in the surface domain structure.

The four-domain structure presented in Figs. 3.16c–e can be considered as a specific magnetic vortex [10]. Rotation of the magnetization by 360° appears in this vortex. Under the action of an axial magnetic field, the vortex moves compactly in the surface of the wire taking part in this way in the magnetization reversal. In Fig. 3.16e, the domain structure is presented, in which the "center" of the vortex is demonstrated. It is a point where the "black" domain wall is changed by the "white" one. The formation of this structure could be considered as the result of the relation between the surface circular anisotropy, external axial magnetic field and domain wall mobility. It is possible to suppose two limit ways of magnetization reversal. The first one is only the rotation of magnetization, when the mobility of domain wall is low enough (the pining is high enough). The second one is the nucleation of the domains with the magnetization axially directed and the successive domain walls motion. This is possible when the circular anisotropy is low and the domain wall mobility is high. In the present experiments, we can observe some intermediate regime, when the rotation of the magnetization is changed in some moment by the domain nucleation and the domain wall motion. The vortex structure appears at this moment.

One of the additional reasons for vortex structure appearance could be the shape anisotropy of the wire. As it was shown in Ref. [10], the non-planar nature of the sample could initiate the formation of magnetization fluctuation of vortex type. From another perspective, in Ref. [11] it was demonstrated theoretically that the formation of some twisted structure in the inner core of Co-rich amorphous wires is possible. Thus, the experimentally observed vortex-type structure could be considered as a reflection of the domain structure rearrangement in the inner core in a context of the strong relation between the domain structure in the inner core and the outer shell of the wire [12, 13].

Therefore, it was found that in the presence of an axial magnetic field, magnetization reversal appears as a fluent rotation of the magnetization followed by the formation of a domain structure containing domains of four different types and curved domain walls. This structure, which can move along the wire surface, could be considered as a magnetic vortex. The formation of the vortex-type structure in the surface of the wire could be related to some twisting process appearing in the inner core of the wire and to the cylindrical-shape anisotropy.

References

1. Humphrey FB, Mohri K, Yamasaki J, Kawamura H, Malmhall R, and Ogasawara I (1987), in *Magnetic Properties of Amorphous Metals* (Eds. Hernando A, Madurga V, Sanchez MC, and Vazquez M). Elsevier Science Publishers, Amsterdam, Netherlands.

2. Vazquez M and Chen DX (1995), *IEEE Trans. Magn.*, **31**, 1229.

3. Usov N, Antonov A, Dykhne A, and Lagar'kov A (1997), *J. Magn. Magn. Mat.*, **174**, 127.

4. Chen DX, Pascual L, Castano FJ, Vazquez M, and Hernando A (2001), *IEEE Trans. Magn.*, **37**, 994.

5. Chizhik A, Zhukov A, Blanco JM, and J. Gonzalez J (2002), *J. Magn. Magn. Mater.*, **249**, 27.

6. Hernando A and Barandiaran JM (1998), *J. Phys. D : Appl. Phys.*, **11**, 1539.

7. Chizhik A, Zhukov A, Blanco JM, and Gonzalez J (2001), *Phys. B*, **299**, 314.

8. Chizhik A, Zhukov A, Blanco JM, and Gonzalez J (2002), *Phys. Status Solidi (a)*, **189**, 625.

9. Chizhik A, Gonzalez J, Zhukov A, and Blanco JM (2002), *J. Appl. Phys.*, **91**, 537.

10. Mohri K, Humphry FB, Kawashima K, Kimura K, and Mizutani M (1990), *IEEE Trans. Mag.*, **26**, 1789.

11. Federici F and Gunn JMF (2001), *J. Magn. Magn. Mater.*, **226–230**, 1607.

12. Zhukova V, Usov NA, Zhukov A, and Gonzalez J, (2002), *Phys. Rev. B*, **65**, 134407.

13. Chizhik A, Gonzalez J, Yamasaki J, Zhukov A., and Blanco JM (2004), *J. Appl. Phys.*, **95**, 2933.

Chapter 4

Interaction between Glass-Covered Microwires

4.1 Introduction

Amorphous bistable microwires have attracted much interest by their technological application as sensors [1] and codification system [2]. Taking into account that for codification devices, a set of the wires is involved, the study of the magnetization reversal of multi-wire system becomes important. The use of wires as the base of sensing elements demands the investigation of some peculiarities on the magnetization reversal at different frequency of the external magnetic field. This chapter is devoted to the studies of the magnetization process in a system containing different number of the glass-covered Fe-rich and Co-rich amorphous wires for conditions of changes of frequency of the magnetic field and the distance among the wires. In particular, we investigated Co-rich microwires with so-called circular bistability [3].

Magnetic Microwires: A Magneto-Optical Study
Alexander Chizhik and Julian Gonzalez
Copyright © 2014 Pan Stanford Publishing Pte. Ltd.
ISBN 978-981-4411-25-7 (Hardcover), 978-981-4411-26-4 (eBook)
www.panstanford.com

4.2 Fe-Rich Microwires

The investigations were performed on Fe-rich glass–coated amorphous microwires of $Fe_{65}Si_{15}B_{15}C_5$ composition (samples with metallic nucleus diameter of 12.6 μm, total diameter 20 μm, and length 25 mm).

Axial hysteresis loops were measured by conventional SQUID and fluxmetric magnetometers and magneto-optical Kerr effect loop tracer. During SQUID experiments, the DC magnetic field has been used. Fluxmetric and longitudinal Kerr effect experiments have been performed under AC magnetic field with frequency up to 1300 Hz. A magnetic field was applied along the axis of the wire and simultaneously in the plane of the light. The rotation of the polarization plane of the reflected light was proportional to the change of the parts of the magnetization, which was parallel to the field-light plane. The measurements have been performed in a single microwire and in a set of several microwires (2, 5, and 10). In the case when several samples were studied, the wires have been placed closely in the same plane on a non-magnetic substrate.

The results of SQUID magnetometer measurements are presented in Fig. 4.1. As expected [4–7], the hysteresis loop for one wire (Fig. 4.1a) has the square shape that is associated with a single large Barkhausen jump for the magnetization process. The increase in the number of the wires causes the increase in the number of jumps. The magnetic field of the first jump decreases (and changes the sign), when the magnetic field of the last jump increases. The value of the magnetic field of the last jump for 10 wires (Fig. 4.1d) exceeds the magnetic field of the jump for one wire by five times. This effect is connected to the mutual interaction in the multi-wire system. As a consequence of the demagnetization fields, each one of the wires could be affected by the additional magnetic field. The superposition of the external and the demagnetizing fields evokes the magnetization reversal in one of the samples when the external field is less than in the case of a single wire. The successive magnetization reversal in the wires changes the magnetic distribution in the multi-wire system and the jumps take place in the superposition of the demagnetizing fields of the remagnetized and the non-remagnetized wires. The shape of the hysteresis loop for a 10-wire system (Fig. 4.1d) testifies that each wire in this system is influenced by the demagnetizing fields of the all surrounding wires.

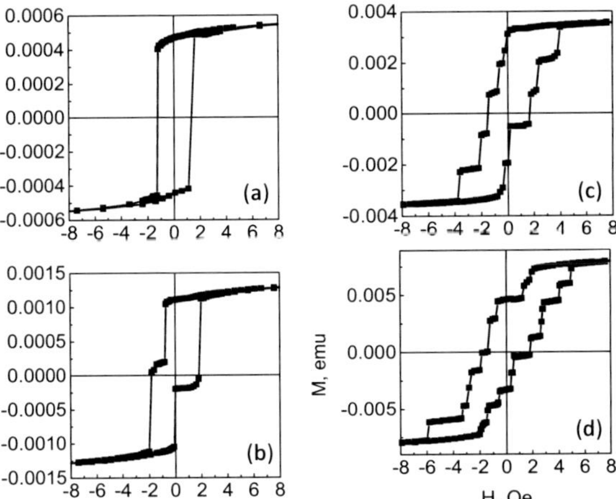

Figure 4.1   Hysteresis loops of the glass-coated $Fe_{65}Si_{15}B_{15}C_5$ amorphous microwire measured by SQUID magnetometer for (a) 1, (b) 2, (c) 5, and (d) 10 microwires.

Figure 4.2 demonstrates the magnetization reversal curves obtained in the system of two wires by the fluxmetric (Fig. 4.2a) and Kerr effect (Fig. 4.2b) methods. The magnetic field was swiped with the frequency of 50 Hz. The two jumps on the fluxmetric loop reflect two large Barkhausen jumps in the volume of two wires. The jumps are not such sharp as in SQUID experiments. This effect is related to frequency dependence of the switching field. The jumps in the Kerr loop are sharp enough but switching field is smaller than in fluxmetric experiments. First, taking into account that the Kerr effect results contain the information about magnetization reversal in surface area of the wire, it is possible to conclude that a large Barkhausen jump takes place in the outer shell of the wire as in the inner core. Second, the sharp shape of the jumps and different values of switching field are also related to the "surface" character of the Kerr effect. The Kerr effect loop reflects the magnetization reversal in small part of the sample.

This process takes lesser time than in the fluxmetric experiments because the fluxmetric loop reflects the magnetization reversal in whole sample volume. Therefore, the jumps are sharper. The difference of switching field in the surface and the volume loops are explained by the fact that the remagnetization process begins in the surface at lower field than in the volume.

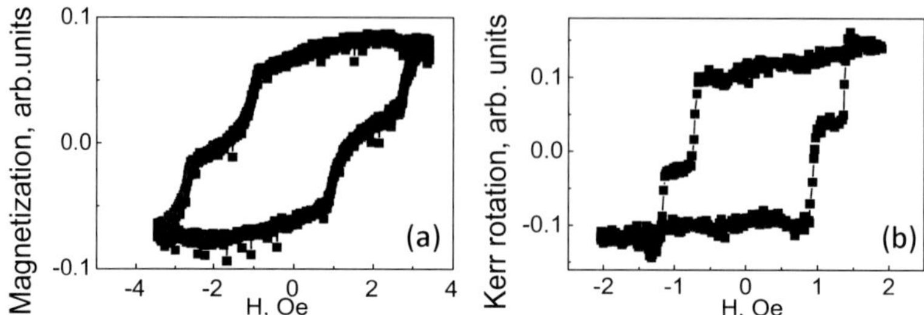

Figure 4.2   Hysteresis loops measured by fluxmetric magnetometer (a) and magneto-optical Kerr effect loop tracer (b) in two-microwire system. Frequency of the magnetic field is 50 Hz.

Frequency dependence of the switching field has been studied using the Kerr effect technique for the two-wire system. The results of the experiments are presented in Fig. 4.3. H_1 and H_2 are the switching fields in the first and the second wire, respectively. An increase in H_1 and H_2 is observed as frequency increases. The difference between H_1 and H_2 decreases and disappears at the frequency about 1200 Hz. Frequency dependencies of the switching field have been analyzed using the calculations presented in [8]. As the result of the solution to the equation of domain wall motion, the following relation has been obtained:

$$H_{cd} = H_{co} + 4fH_0(L + 2I_sA)/K,\qquad(4.1)$$

where H_{cd} is the dynamic switching field, H_{co} is the static switching field, f is the frequency of the magnetic field, H_0 is the amplitude of the magnetic field, L is the damping coefficient, I_s is the saturation magnetization, K is the elastic coefficient, and A is a proportionality constant. The experimental results are in good agreement with the theoretically obtained linear dependence of the switching field on the frequency for H_1 and H_2 (see Fig. 4.3). This dependence could be described in terms of domain walls pinning mechanism. The nucleation process and successive domain walls motion are associated to the overcoming of the energy barrier. The thermoactivation mechanism of the overcoming was satisfactorily employed for the explanation of the switching field fluctuations observed in magnetic materials with bistable properties. The frequency of the applied field in this way affects the value of the switching field.

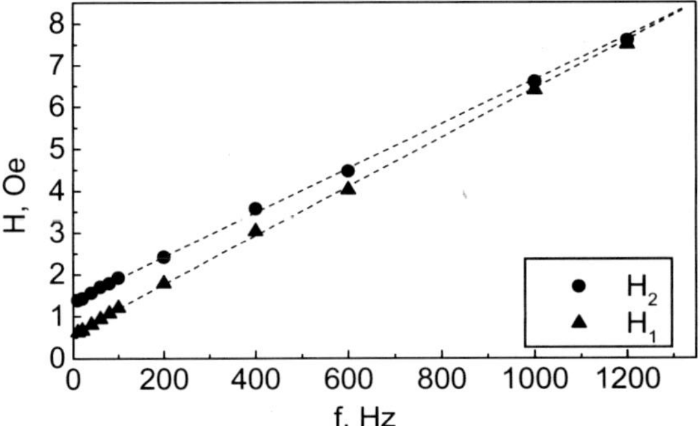

Figure 4.3   Dependence of the difference between the two jumps on the frequency of the magnetic field in the two-microwire system.

The decrease of ΔH (difference between the two Barkhausen jumps) could be related to the finite character of the magnetization reversal in the wire, namely the finite value of the domain wall velocity. As was noted above, different values of the magnetic field of the two jumps could be ascribed to the influence of the demagnetizing field on the magnetization reversal. In particular, the second jump is happened in the superposition of the external field and the demagnetizing field of the wire, which is first re-magnetized. There are two time-dependent processes: the arising of the external magnetic field and the formation of the demagnetizing field during the magnetization reversal. When the values of the time of these two processes are close, the influence of the demagnetizing field decreases and accordingly ΔH decreases. The limit situation takes place at high frequency when there is no difference between the jumps. The field value necessary for the magnetization reversal in the second wire is reached considerably earlier than the magnetization reversal is finished in the first wire. The distribution of demagnetizing field around the first wire is not formed yet and, therefore, there is no influence of the re-magnetized wire on the magnetization reversal in the second wire.

The dependence of ΔH differences between the two jumps for the two-wire system has been measured by the conventional fluxmetric method (Fig. 4.4). The decrease of the ΔH differences with the distance between the wires is associated with the decrease of the demagnetizing field. The performed experiments show that a mutual influence is displayed up to the distance of 2.5 mm between the wires. For a distance longer than 2.5 mm, the magnetization reversal appears independently in the wires. The experimental results have been approximated by the dependence $\sim 1/X^3$ (line in Fig. 4.4), taking into account such dependence of the demagnetizing field value on the distance from the magnetized sample.

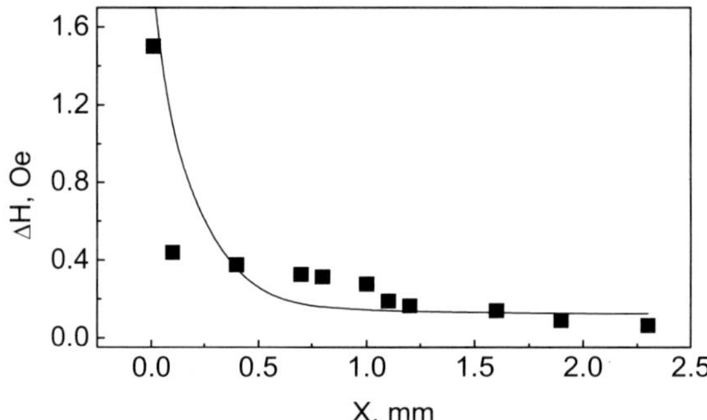

Figure 4.4   Dependence of the difference between the two jumps on the distance between the wires in the two-microwire system.

This description of the experimental dates by the above dependence suggests that the influence of the demagnetizing field is the main reason for the observed successive jumps in the multi-wire system.

The mutual influence of the demagnetizing field of the wires is reflected in the volume and in the surface hysteresis loops. The influence of the frequency of the applied magnetic field on the magnetization reversal process has been examined in the supposition of the pinned domain walls motion. In addition, the transformation of the hysteresis loop with the increase of the distance between wires has been studied. The obtained results are in agreement with the model of the mutual influence of the wires.

4.3   Co-Rich Microwires

The investigations were performed on two glass-covered amorphous microwires which are closely placed (nominal composition

$Co_{67}Fe_{3.85}Ni_{1.45}B_{11.5}Si_{14.5}Mo_{1.7}$, metallic nucleus radius R = 11.2 µm, glass coating thickness T = 0.2 µm). The length of the studied microwires was 5 cm. The distance between the microwires was set equal to the double thickness of the glass coating (0.4 µm). The experiments have been carried out using the transverse magneto-optical Kerr effect in circular and axial magnetic field. An AC electric current flowing through the wires produced the AC circular magnetic field. The frequency of the current was 50 Hz.

The intensity of the light reflected from wires was proportional to the magnetization perpendicularly oriented to the light plane. Therefore, the light intensity was proportional to the circular projection of the magnetization. Taking into account that the circular magnetic bistability appears in circular magnetic field [3], which can be produced by passing an electric current through the wire, we used a special experimental configuration in which the electric current flows through the wires in opposite directions (Fig. 4.5). Such a configuration has been previously used for fluxgate sensor applications [9, 10]. In our experiments, the microwires are placed in a specific magnetic field configuration, which cannot be realized in conventional axial magnetic field experiments. This configuration was not tried before for glass-covered microwires. The application of the Kerr effect is the unique possibility to get information about the magnetization reversal in this configuration.

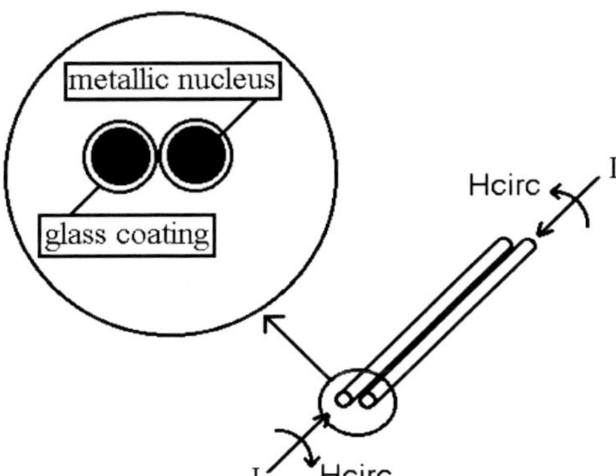

Figure 4.5   Schematic drawing of experimental configuration.

For the present work, we have chosen a system of two Co-rich microwires because of the circular magnetic bistability effect observed in such wires [3]. This effect is related to the large Barkhausen jump between two states with opposite directions of the circular magnetization in the microwire outer shell. Taking into account that the Kerr intensity, I, is proportional to the transverse magnetization, M, in the wire surface area, it is possible to consider that $I/I_S = M/M_S$, where I_S is the intensity of the Kerr signal in the saturation state and M_S is the transverse saturation magnetization. Therefore, the perfect rectangular transverse Kerr effect hysteresis loop should be associated with the circular bistability effect. Figure 4.6 shows such a hysteresis loop obtained for a single Co-rich microwire [3]. The insets in Fig. 4.6 show two states with opposite circular magnetizations, between which the large Barkhausen jump appears.

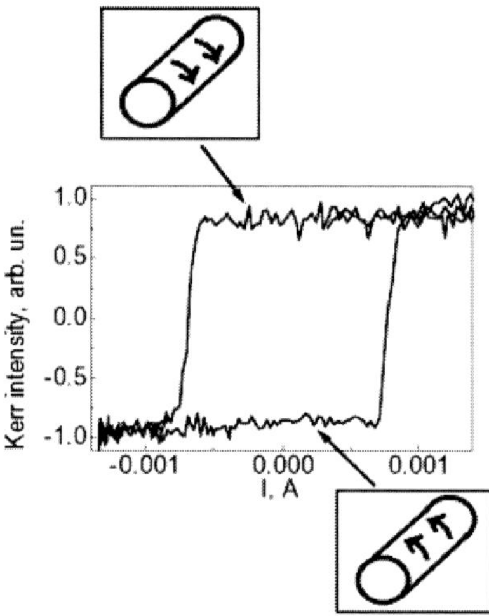

Figure 4.6   Transverse Kerr effect dependence on AC electric current flowing through the wire for one-wire configuration. The direction of the circular magnetization in the wire outer shell is shown in insets.

Figure 4.7c presents the dependence of the transverse Kerr effect on the electric current flowing through two wires in the absence of the DC axial magnetic field. As for the case of axial bistability [5, 6], we consider that the magneto-optical signal from the two–microwire system (I_{SYS}) is given by the superposition of the signals coming from the two wires separately considered ($I_{SYS} = I_1 + I_2$). Because of the electric current that flows through the two wires in opposite directions, the circular magnetic fields induced in the two wires have also opposite directions. Consequently, the Kerr effect hysteresis loop in the second wire is "reversed" with respect to the hysteresis loop in the first wire. Therefore, the complete disappearance of the total Kerr signal from two microwires should be expected. However, the presence of two peaks on the hysteresis loop is observed (Fig. 4.7c). The two jumps "up" and two jumps "down" are ascribed to successive jumps of the circular magnetization in the two microwires.

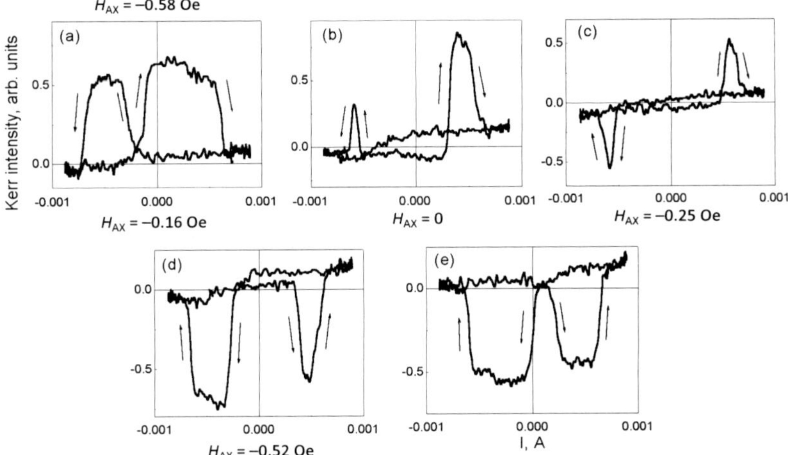

Figure 4.7 Transverse Kerr effect dependencies on the AC electric current flowing through two wires in the presence (a, b, d, e) or the absence (c) of DC axial magnetic field. The absolute values of the DC axial magnetic field are shown in the figure.

In the presence of the DC axial magnetic field (Figs. 4.7a,b,e) the dramatic transformation of the hysteresis loop is observed. The height and the position of the peaks change with the increase of the DC magnetic field. The shape of the hysteresis loop depends also on the sign of the DC axial magnetic field. When the DC magnetic field is high enough, two almost separated hysteresis loops with different values of the AC switching field are observed (Figs. 4.7a,e).

The main explanation for the obtained experimental results is the influence of the axial magnetic field, i.e., external field or stray field created by the neighboring wire, on the circular magnetization reversal [11]. As against the case of axial bistability, this influence is not so obvious, especially taking into account that the axial magnetic field of the determined direction assists the nucleation of circular domain of the determined direction.

The analysis of experimental hysteresis loops has been performed under the consideration of the mutual interaction of the wires and taking into account a spontaneous character of the magnetization reversal in bistable microwires. Based on our previous results related to the mutual interaction of the axially bistable microwires [6], we could suppose that the value of the stray field on the distance of 0.4 μm (double thickness of glass coating) should be high enough. As a consequence of the stray fields produced by the inner cores of the two microwires, each of them is additionally affected by the axial magnetic field as well as by the circular magnetic field produced by the electric current flowing through the wires. The superposition of the circular magnetic field and the axial stray field causes the reversal of the circular magnetization in one of the wires when the circular field is less than in the case of the single wire. The direction of the axial magnetization in the inner core changes to the opposite one during the magnetization reversal. After the magnetization reversal in the first wire, the spatial magnetic field distribution changes. The direction of the stray field of the first wire is changing into the opposite. The superposition of the circular field and the stray field with opposite direction causes the magnetization reversal of the second wire at larger circular fields in comparison with the one corresponding to the single wire. Consequently, two circular Barkhausen jumps appear in the system of two microwires with circular magnetic bistability, when the electric current changes from −0.001 A to +0.001 A.

The modeling of the magnetization reversal in the two-wire system has been performed considering that the values of the switching current and the circular switching field, which are similar for two single wires separately considered, are different when the wires compose the two-wire system with mutual interactions (Figs. 4.8c,h). Thus, the two successive jumps of circular magnetization are observed, when the circular fields increases from "−" to "+," and two jumps when the circular field decreases from "+" to "−." Since the direction of the circular magnetic field in the two wires is opposite, the Kerr effect hysteresis loop in the second wire is "reversed" (Fig. 4.8h) with respect to the hysteresis loop in the first wire. To get the model picture for magnetization reversal in the two-wire system, the hysteresis loop of the first wire (Fig. 4.8c) is superposed to the hysteresis loop of the second wire (Fig. 4.8h). In this way, the hysteresis loop of the two-wire system looks as presented in Fig. 4.8m: two jumps "up" and two jumps "down" with respect to "zero" background level. It is worth to mention that as in the experimental hysteresis loop (Fig. 4.7c), the absolute disappearance of the total model hysteresis is not observed because of the mutual influence of the wires: the magnetization reversal in the two wires takes place at different values of the circular magnetic field.

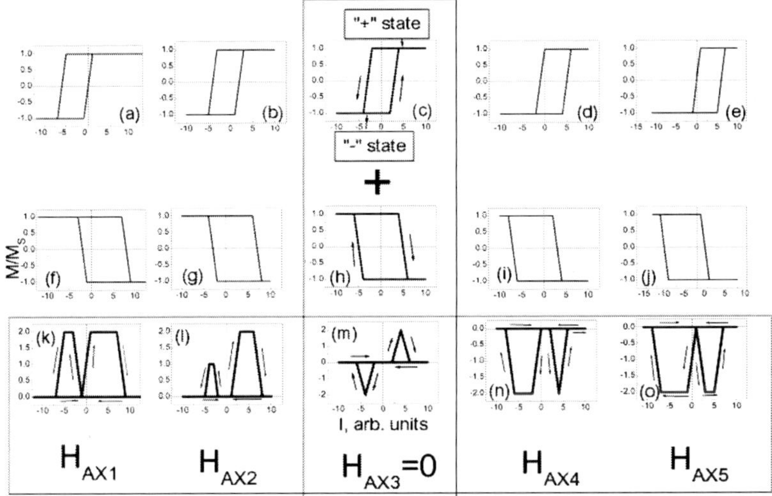

Figure 4.8 Modeling of the magnetization reversal in two-wire system. (a–e) Magnetization reversal in the "first" wire; (f–j) magnetization reversal in the "second" wire; (k–o) magnetization reversal in the two–wire system. $H_{AX1} < H_{AX2} < H_{AX3} < H_{AX4} < H_{AX5} < H_{AX3} = 0$.

The modeling of the magnetization reversal in the presence of the DC axial magnetic field is also presented in Fig. 4.8. The model hysteresis loops for the first microwire (Figs. 4.8a–e) were superposed to the model hysteresis loops for the second microwire (Figs. 4.8f–j). The result of the superposition is presented in Figs. 4.8k–o. The effect of the DC axial magnetic field is associated with its influence on the nucleation of the circular domains. The reason for the shift of the transverse hysteresis loop is that the DC axial magnetic field assists the nucleation of the circular domain of one type (for example "+") and inhibits the nucleation of the domain of the other type ("–"). The shift of the transverse hysteresis loop increases with the increase of the value of the DC axial magnetic field. When the axial magnetic field is reversed, the hysteresis loop shifts to the other direction. Because of the opposite directions of circular magnetization in the two microwires during the magnetization reversal, the shifts in the two wires appear in different directions (see, for example, Figs. 4.8d,i). The modeling was performed under the conditions of $H_{AX1} < H_{AX2} < H_{AX3} < H_{AX4} < H_{AX5}$ and $H_{AX3} = 0$. The results of the modeling for the case of the application of the axial field are presented in Figs. 4.8k, l,n,o. If the axial field is high enough, the model curves exhibit the almost separated bistable hysteresis loops in the two wires (Figs. 4.8k,o) as in the experimental curves (Figs. 4.7a,e).

Magneto-optical Kerr effect investigations have been carried out in the system of two Co-rich glass–covered amorphous microwires with circular bistability. The experimental configuration, when the electric current flows through the two wires in opposite directions and produces circular magnetic fields, has been used for the first time for a system of two glass-covered microwires. The shape of the obtained hysteresis loops reflects the interaction of the microwires in the conditions of the existence of circular bistability. It was found that the magnetization reversal in the system of two Co-rich microwires differs from the magnetization reversal in the single Co-rich microwire.

The analysis of the experimental results has been performed taking into account the influence of the axial magnetic field (external field or stray field of neighboring wire) on the circular magnetization reversal in the single microwire. The very good agreement of the experimental results (Fig. 4.7) and the results of the modeling (Figs. 4.8k–o) allows us to conclude that the main assumptions of our model are adequate. The superposition of the circular magnetic field produced by electric current flowing through the wires in opposite directions and axial stray fields produced by the inner cores of the two microwires creates the spatial distribution of the magnetic field. Owing to this distribution, in two microwires with circular bistability, which are placed closely, the successive opposite jumps of circular magnetization appear. The application of the DC axial magnetic field results in the transformation of the hysteresis loop, which is related to the effect of the axial field on the circular domains nucleation process. The results of the modeling permit us to conclude that the value of the stray fields produced by the inner core of microwires is high enough to change the magnetization reversal process in the circularly magnetized outer shell, when the distance between the two microwires is equal to the double thickness of the glass covering.

References

1. Vazquez M and Hernando A (1996), *J. Phys. D*, **29**, 1.

2. Zhukov A, Gonzalez J, Blanco JM, Vazquez M, and Larin V (2000), *J. Mater. Res.*, **15**, 2107.

3. Chizhik A, Gonzalez J, Zhukov A, and Blanco J M (2003) *J. Phys. D*, **36**, 419.

4. Vazquez M and Chen DX (1995), *IEEE Trans. Magn.*, **31**, 1229.

5. Squire PT, Atkinson D, and Atalay S, (1995), *IEEE Trans. Magn.*, **31**, 1239.

6. Chizhik A, Zhukov A, Blanco JM, and Gonzalez J, (2001) *J. Non Crystall. Solids,* **287**, 374.

7. Chizhik A, Zhukov A, Blanco JM, Szymczak R, and Gonzalez J (2002), *J. Magn. Magn. Mater.*, **249**, 99.

8. Zhukov A, Vazquez M, Velazquez J, Garcia C, Valenzuela R, and Ponomarev B (1997), *Mater. Sci. Eng.*, **226–228**, 753.

9. Nielsen O V, Gutierrez J, Hernando B, and Savage HT (1990), *IEEE Trans. Magn.*, **26**, 276.

10. Primdahl F (1979), *J. Phys. E: Sci. Instrum.*, **12**, 241.

11. Chizhik A, Gonzalez J, Zhukov A, and Blanco JM (2002), *J. Appl. Phys.,* **91**, 537.

Chapter 5

Circular Magnetic Bistability in Co-Rich Amorphous Microwires

5.1 Introduction

Circular magnetic bistability associated with a large Barkhausen jump between two states with opposite directions of circular magnetization has been observed by magneto-optical Kerr effect in nearly zero magnetostrictive glass-covered Co-rich amorphous microwire. The influence of external axial tensile stress on the circular switching field has been studied. Observed results have been related with a circular magneto-elastic anisotropy induced by a tensile stress in the outer transversally magnetized shell of the microwire. The coating introduces additional internal stresses due to the difference between the thermal expansion coefficients of glass coating and metallic nucleus. Therefore, microwires of the same composition can show different magnetic properties because of the different magnetoelastic energy. The investigation of GMI in $Co_{67}Fe_{3.85}Ni_{1.45}B_{11.5}Si_{14.5}Mo_{1.7}$ microwires with geometric ratio

Magnetic Microwires: A Magneto-Optical Study
Alexander Chizhik and Julian Gonzalez
Copyright © 2014 Pan Stanford Publishing Pte. Ltd.
ISBN 978-981-4411-25-7 (Hardcover), 978-981-4411-26-4 (eBook)
www.panstanford.com

ρ between 0.69 and 0.98 [1] showed that the microwire with very thin glass covering demonstrated very high GMI ratio (around 615%).

5.2 Circular Magnetic Bistability Effect Related with a Large Circular Barkhausen Jump

Magneto-optical studies of glass-covered amorphous microwires of nominal composition $Co_{67}Fe_{3.85}Ni_{1.45}B_{11.5}Si_{14.5}Mo_{1.7}$ (metallic nucleus radius R = 11.2 µm, glass coating thickness T = 0.2 µm) have been performed. During experiments, a mechanical load was added to microwires with the purpose to apply an axial tensile stress on them [2].

Figure 5.1 presents the transverse Kerr effect dependence on the AC electric current I flowing with the frequency of 50 Hz along the sample with an external tensile stress σ as a parameter. The values of external tensile stress are marked in the figure. In the absence of external tensile stress (Fig. 5.1a), the shape of the hysteresis loop is perfectly rectangular with sharp vertical areas associated with quick enough reversal of circular magnetization. These reversals could be considered as a realization of circular magnetic bistability in the form of large Barkhausen jumps between two states with opposite directions of circular magnetization. The application of external tensile stress causes the change of switching field, H_{SW} (associated with the switching current). In the present experiments, the circular switching field should be considered as the field at which the large Barkhausen jump stars.

Also the change of the I_R/I_S ratio under the tensile stress is observed (I_R is the intensity of the Kerr signal in the remanent state and I_S is the intensity of the Kerr signal in the saturation state). Taking into account that the Kerr intensity is proportional to transverse magnetization M in the surface area of the wire, it is possible to consider that $I_R/I_S = M_R/M_S$, where "M_R is the transverse remanent magnetization and M_S is the transverse saturation magnetization."

Figure 5.2 shows the external tensile stress dependence of H_{SW} and M_R/M_S ratio. The H_{SW} values have been obtained from switching current values using the formula for circumferential field

$H_{\mathrm{CIRC}} = Ir/2\pi R^2$. The calculation has been performed for the surface of the metallic nucleus, i.e., for $r = R$. The circular switching field decreases with the stress σ showing a minimum for around 50 MPa followed by an increase of H_{SW} at higher value of stress. The $M_{\mathrm{R}}/M_{\mathrm{S}}$ ratio increases as tensile stress increases and reaches the saturation at the stress value of around 70 MPa.

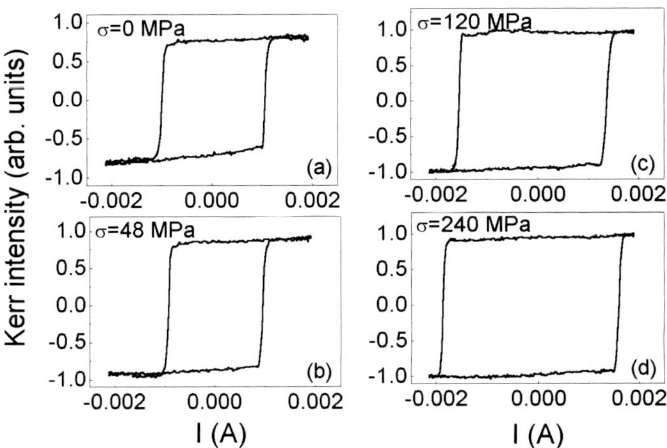

Figure 5.1 Transverse Kerr effect dependence on AC electric current flowing with the frequency of 50 Hz through the wire for different tensile stresses: (a) 0 MPa, (b) 48 MPa, (c) 120 MPa, and (d) 240 MPa.

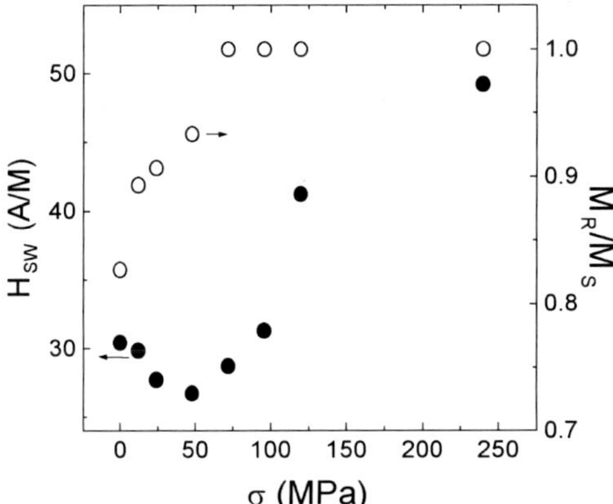

Figure 5.2 External tensile stress dependence of circular switching field and circular remanent magnetization to saturation magnetization ratio.

To study in details the magnetization reversal process, which determines the circular bistability effect we performed Kerr effect experiments under axially applied magnetic fields. Figure 5.3 presents the transverse Kerr effect dependence on the AC axial magnetic field of the frequency of 50 Hz with the DC electric current I_{DC} as a parameter. Let us consider first the hysteresis loop in the absence of I_{DC} (Fig. 5.3d). When the external axial magnetic field H_{AX} increases from −4 Oe, a monotonic increase of Kerr signal is observed. This increase is related to the rotation of the magnetization from axial to circular direction in the outer shell of the wire. After that, the sharp jump of the signal with the change of the signal sign takes place.

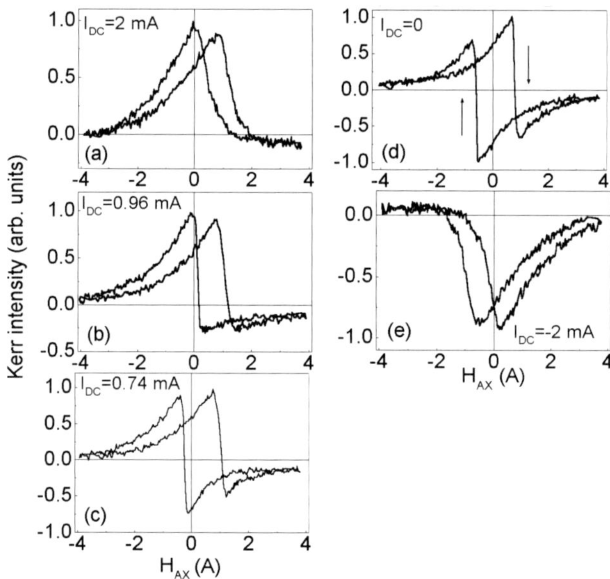

Figure 5.3 Transverse Kerr effect dependence on the AC axial magnetic field of the frequency of 50 Hz with in the presence of DC electric current.

This is related to the nucleation of new circular domain. It is worth mentioning that the sign of the nucleated circular domain is associated with the sign of the external magnetic field. In the presence of DC current, a shift of the peaks positions and change of the peaks values are observed. This evolution (Figs. 5.3b,c) shows how the DC current changes the conditions of the circular domain nucleation. The DC electric current makes favorable the nucleation of circular domain with the corresponding magnetization that is reflected in the shift of the peaks. The decrease of the peaks values with their subsequent disappearance demonstrates that the new domain forms from the state with magnetization tilted from the axial direction and finally exactly from the axially magnetized state. When the electric current is high enough (Figs. 5.3a,e), only one domain exists and the rotation of magnetization is observed.

In order to study the details of circular domain nucleation, we have performed Kerr effect experiments in presence of both AC and DC electric currents (Fig. 5.4). The magnetization reversal has been investigated under 120 MPa tensile stresses. The solid line in Fig. 5.4 demonstrates the rectangular hysteresis, which is related to stable circular Barkhausen jumps observed in the absence of DC electric current (I_{DC} = 0). Application of DC electric current causes in common way the disappearance of circular bistability because the DC circular magnetic field inhibits the nucleation of the circular domain of one type. However, in the very narrow interval of DC electric current, the non-stable hysteresis loops exist. They are presented in Fig. 5.4 by lines with symbols. The amplitude of these hysteresis loops depends on the value of I_{DC} and decreases as the DC electric current increase. These curves demonstrate the existence of an intermediate state in the way of the formation of the circular bistability. The plane areas, which take place in these hysteresis loops, can be related to the existence (in metastable condition) of the state with the magnetization tilted from the circular direction or to the nucleation of the reversed circular domain. The degree of the tilting depends on the I_{DC}. The origin of this tilting could be related to weak helicoidal magnetic anisotropy. The domain structure observation (Kerr effect microscopy) can clarify the origin of these hysteresis loops, domain wall propagation details and contribution of domain wall damping mechanisms.

The existence of the domain structure with circular orientation of the magnetization in the outer shell of the studied microwire allows us to suppose that the sign of the magnetostrictive coefficient λ_S should be negative. To check this fact, we used the small-angle magnetization rotation (SAMR) method [3] to evaluate the magnetostriction coefficient. In this method, the sample is saturated by an axial magnetic field, while applying simultaneously a small AC transverse field, created by an AC electric current of frequency ω flowing along the sample. The combination of these fields leads to a reversible rotation of the magnetization within a small angle out of the axial direction. The induction voltage, $V(2\omega)$, due to the magnetization rotation is detected by a coil wound around the microwire. The magnetostriction coefficient is determined from the dependence of $V(2\omega)$ on applied stress and bias field. It was found that $\lambda_S = -2.12 \times 10^{-7}$.

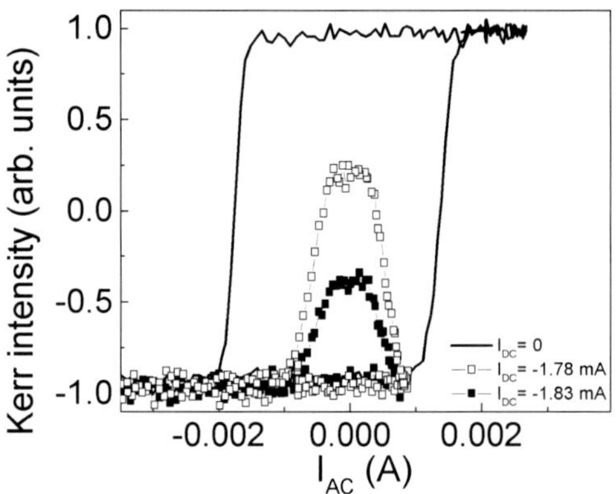

Figure 5.4   Transverse Kerr effect dependence on AC electric current flowing through the wire in the presence of DC electric current.

The circular magnetic bistability, as the classical longitudinal magnetic bistability [4, 5], is related to the magnetoelastic anisotropy in the circular direction. The circular magnetoelastic anisotropy results in the appearance of circular mono-domain structure in the outer shell. It is known that in negative magnetostrictive wires, the axial tensile stress produces circular anisotropy in the outer shell. In the presented experiments, this effect is reflected in the increase of circular remanence magnetization and in the increase of the squareness of the hysteresis loops in presence of tensile stress. The experimental results have been analyzed taking into account the strong influence of the magnetostriction on the domain structure in the amorphous wires. The circular bistability effect is related to the formation of circular domain walls. The circular switching field should be proportional to the energy γ required to form the circular domain wall [6, 7].

The effect of circular magnetic bistability discovered in glass-covered microwire having negative magnetostrictive constant is related with a large Barkhausen jump between two states with opposite directions of circular magnetization. The influence of the applied tensile stress on the circular switching field and the circular remanent magnetization can be explained considering the existence of circular magnetoelastic anisotropy in the outer shell of microwire, which is produced by internal and external tensile stresses. The existence of metastable states with the magnetization tilted from circular direction has been found as an intermediate condition between two stable circularly magnetized states.

5.3 Circular Magnetic Bistability Induced by Tensile Stress

Magneto-optical studies of glass-covered amorphous microwires of nominal composition $Co_{67}Fe_{3.85}Ni_{1.45}B_{11.5}Si_{14.5}Mo_{1.7}$ (metallic nucleus radius 11.2 μm, glass coating thickness 3 μm) have been performed. The transverse hysteresis loop measured using transverse Kerr effect with AC electric current flowing along the wire under application of the tensile stress σ is presented in Fig. 5.5. The observed change of the shape of the loop is related to the formation of the circular domain in the outer shell of the microwire. The monotonic hysteresis loop is observed with $\sigma = 0$. The sharp jumps of magneto-optical signal arising in the presence of the tensile stress should be attributed to so-called circular bistability. This circular bistability appeared as a result of a large Barkhausen jump of the circularly oriented magnetization, which has been found in Co-rich microwires with negative magnetostriction. Another significant feature is the increase of the value of the switching current (and accordingly, the circular switching field) with increasing tensile stress. This indicates the change of coercivity of circular domain structure under the tensile stress.

The experiments performed in AC axial magnetic field (Fig. 5.6) permitted us to clarify the details of the tensile induced transformation of the domain structure and the features of the magnetization reversal. First, the longitudinal Kerr effect dependence on axial magnetic field in the absence of tensile stress (Fig. 5.6a) exhibits perfectly squared shape that could be associated with a large Barkhausen jump of the axial magnetization in the outer shell of the microwire.

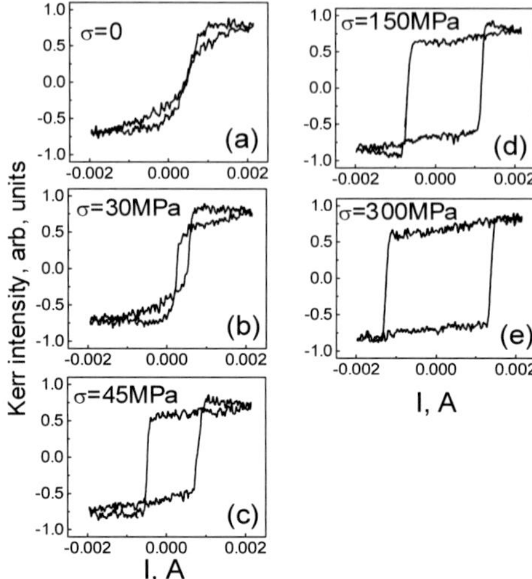

Figure 5.5   Transverse Kerr effect dependencies on the electric current flowing through the wire for different tensile stresses: (a) 0 MPa, (b) 30 MPa, (c) 45 MPa, (d) 150 MPa, and (e) 300 MPa.

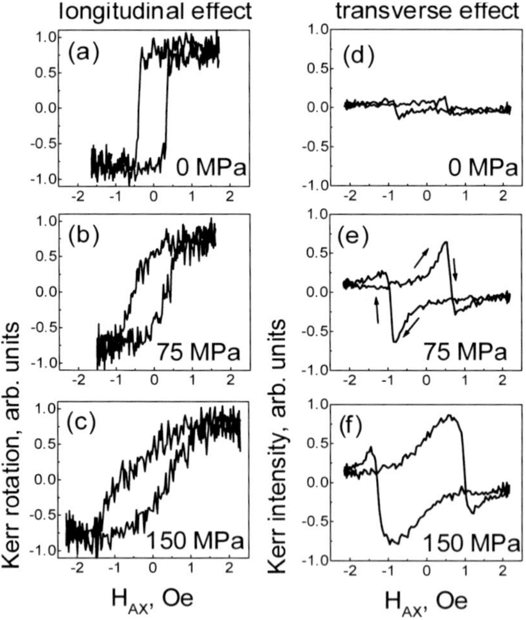

Figure 5.6   Longitudinal (a–c) and transverse (d–f) Kerr effect dependencies on the axial magnetic field in the presence of tensile stress: (a, d) 0 MPa, (b, e) 150 MPa, (c, f) 300 MPa.

The comparison of Figs. 5.5a and 5.6a permits us to consider that the so-called maze domain structure exists in the outer shell of the studied wire in the absence of tensile stress. The maze domain structure has closure domains with magnetization, which is aligned parallel to the axis of the wire. The Barkhausen jump of the axially oriented magnetization takes place in the axial AC field (Fig. 5.6a) and the magnetization rotation is observed in the AC circular field (Fig. 5.5a). Application of an axial stress induces drastic changes in the remagnetization process: the hysteresis curve of the longitudinal effect dependence on the axial field loses the squareness (Figs. 5.6b,c) and the hysteresis curve of the transverse effect dependence on the circular field becomes squared (Figs. 5.5b–e). This could reflect the stress-induced rearrangement of the domain structure: from maze type to circular bamboo type.

The transverse Kerr effect experiments in the AC axial field (Figs. 5.6d–f) give additional information about the formation of circular domain structure and open new details of the magnetization reversal process. First, it is worth mentioning that these three graphs were plotted in the same scale of the Kerr intensity. Figures 5.6e,f represent the typical behavior of circular domain structure in the axial field. The monotonic increase of the Kerr signal with increase of field from –2 Oe could be related to the rotation of magnetization from axial to circular direction. Rather sharp jump of the signal, which is followed by the change of the sign of the signal, is associated with the nucleation of new circular domain with the opposite direction of the magnetization. The last part of the loop is the monotonic decrease of the signal reflected the magnetization rotation from circular into the axial direction. Since the axial projection of the magnetization inside the circular domain is absent, the nucleation of a circular domain cannot be observed in the longitudinal curves (Figs. 5.6b,c). In the absence of tensile stress, the magnetization reversal of maze domains creates the weak reflection in the transverse loop with low amplitude (Fig. 5.6d). The existence of this curve could be explained by the small inclination of magnetization from axial direction in vicinity of the closure domains.

The other observed feature is the increase of the absolute value of the transverse magneto-optical signal and the value of coercivity under the tensile stress (Figs. 5.6d–f). This is a result of the re-arrangement of the domain structure related with the appearance of a bamboo domain structure in the outer shell. As should be expected, the increase of the circular magnetization results in the increase of the circular domain coercivity with increasing tensile stress.

Besides, we have performed the experiments by applying an axial DC bias magnetic field in addition to the AC circular magnetic field and applied tensile stress (Fig. 5.7). It has been found that the axial DC field destroys the circular bistability condition and somehow acts "against" the tensile stress. Figure 5.7 shows the transverse Kerr effect hysteresis loops under various axial magnetic fields. Figure 5.7a demonstrates the circular bistability induced by the tensile stress of 150 MPa. There are two features when the axial magnetic field is applied (Figs. 5.7b–d): decrease of absolute value of the jump of the magneto-optical signal and decrease of coercivity.

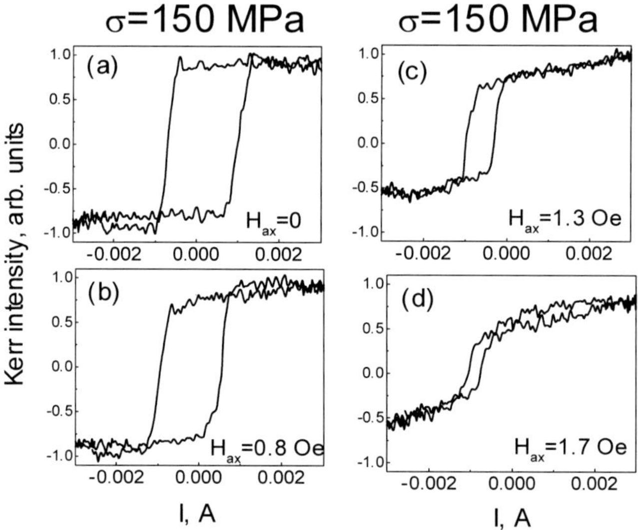

Figure 5.7 Transverse Kerr effect dependencies on the electric current for different value of DC axial magnetic field: (a) 0 Oe, (b) 0.8 Oe, (c) 1.3 Oe, and (d) 1.7 Oe.

These features should be attributed to the inclination of circular magnetization toward the axial direction and accordingly to the decrease of circular projection of the magnetization in the outer shell under the effect of DC axial magnetic field. In this way, the magnetization jumps appear between two states with magnetization successively inclined from circular direction by the DC axial magnetic field. In the magnetic field of 2 Oe the circular bistability effect disappears and only the rotation of magnetization is observed. So, it is possible to conclude that this DC magnetic field "compensates" the influence of the tensile stress of 150 MPa on the domain structure.

The results obtained have been explained taking into account the strong correlation between the magnetostriction and the domain structure in the amorphous microwires. This correlation is based on the competition between the magnetostatic and magnetoelastic energies. The magnetostatic energy depends on the demagnetizing field

$$H_d = NM_S, \tag{5.1}$$

where M_S is the saturation magnetization and N is the demagnetizing factor given for the case of long cylindrical body as

$$N = 4\pi([\ln(2L/D) - 1]\,(L/D)^2), \tag{5.2}$$

where L is the length of the sample and D is the diameter of the sample. While the magnetoelastic energy determines by the magnetostriction value and by both applied, σ_{app}, and internal, σ_{int}, stresses. The corresponding constant of magnetic anisotropy, K_{me}, related with the magnetoelastic contribution can be expressed as

$$K_{me} = (3/2)\lambda_S\sigma, \tag{5.3}$$

where λ_S is the magnetostrictive constant and $\sigma = \sigma_{appl} + \sigma_{int}$. Following Refs. [8, 9], we supposed that the relation between λ_S and σ is determined by the expression

$$\lambda_S(\sigma) = \lambda_S(0) - A\sigma, \tag{5.4}$$

where $\lambda_S(0)$ is the magnetostriction constant without applied stresses and A is the positive coefficient. This expression was experimentally found for conventional wires and ribbons [10]. A relatively small $\lambda_S(0)$ value of about 10^{-7} was estimated during the analysis of the experimental results (like that expected for the investigated Co-rich composition). We consider our experimental results in supposition that one of the reasons of the observed transformation of the hysteresis loops is the change of the sign of the magnetostriction constant [11]. The large Barkhausen jump of the axial magnetization in the outer shell for $\sigma = 0$ (Fig. 5.6a) could be attributed to the positive value of $\lambda_S(\sigma)$, when the large Barkhausen jump of the circular magnetization in the outer shell for $\sigma = 45$ MPa (Fig. 5.5c) could be attributed to the negative value of $\lambda_S(\sigma)$. Considering that $\lambda_S(\sigma) = 0$ for tensile stress of about 30 MPa, we could estimate the value of phenomenological A constant for glass-covered wires is about 3×10^{-9} MPa. This value differs from the value of *A* for conventional wires and ribbons (10^{-10} MPa) [10]. The possible reason for this difference is the additional internal stress introduced by the glass coating and also, maybe as a more relevant feature, that our experimental work involved the surface magnetic behavior different from that of Ref. [10] concerning the bulk magnetic behavior of the amorphous wire and ribbon-shaped alloy.

The transverse hysteresis loop with perfectly squared shape appears step by step under the applied tensile stress (Fig. 5.5). In the intermediate stage (tensile stress of 30 MPa), the Kerr effect curve contains monotonic parts and jumps (Fig. 5.5b), which could be associated with the rotation of magnetization and the motion of domain walls between circular domains. Also, the minor loop is observed at this tensile stress. The existence of such intermediate stage permits us to conclude that the transformation of the domain structure under the tensile stress from maze type to circular single domain appears through the multi-domain circular bamboo structure.

Analyzing our results, we used the core–shell model developed for Co-rich amorphous wires of a finite length [12]. As was shown in Ref. [12], the demagnetizing field of surface magnetic charges can arise near the wire ends for the case of finite length wires. The magnetostatic energy of the magnetic charges could be reduced by twisting of magnetization near the wire ends. Owing to this twisting, the intermediate area with circular and axial projections of magnetization could exist between the inner core and the outer shell. Taking into account the strong correlation between the magnetization in the inner core and the outer shell, the appearance of domain walls between circular domains could be energetically favorable in order to diminish the magnetostatic energy in whole volume of the wire.

When the external stresses, σ_{appl}, are small enough, the stability of the multi-domain circular structure in the nearly zero magnetostrictive composition could be reasonable because of the competition between the magnetostatic and the magnetoelastic energy at the condition, when the magnetoelastic energy is low enough. Successive transformation of hysteresis loop (Figs. 5.5c–e) demonstrates that the multi-domain bamboo structure can exist in the narrow interval of the external stress. The increase of the external tensile stress makes the multi-domain structure non-stable, due to the increase of the domain wall energy

$$\gamma = 2(A_{ex}K)^{1/2}, \tag{5.5}$$

where A_{ex} is the exchange energy and K is the magnetic anisotropy constant, which in the case of amorphous state depends mainly on the magneto-elastic component, K_{me}.

Applying additional external stress, we increase the magnetoelastic energy and, consequently, the domain wall energy. In such conditions, the bamboo domain structure disappears and the single circular domain state is observed (Fig. 5.5c).

The tensile stress–induced circular magnetic bistability as a result of the formation of a circular domain structure has been found. The increase of the tensile stress causes the increase of the value of the circular projection of the magnetization in the outer shell of the microwire that reflects on the coercive properties of the domain structure. The disappearance of the circular bistability under application of axial DC magnetic field can be related to the field-induced inclination of circular magnetization the toward axial direction. The change of the sign of the magnetostriction constant and its small value are crucial for the observation of changes of the type of the domain structure and for its high sensitivity to external stresses.

5.4 Domain Structure in Glass-Covered Co-Rich Microwires in Presence of Tensile Stress

The surface domain structure has been studied in microwires using the Kerr microscope. The microwires with ρ of 0.98, 0.94, and 0.82 demonstrate the circular mono-domain structure in the absence and the presence of the tensile stress. The microwire with ρ of 0.79 shows the circular multi-domain structure of bamboo-type in the absence of the tensile stress (Fig. 5.8a). Tensile stress causes the successive transformation of multi-domain structure to mono-domain one (Figs. 5.8b,c) [13].

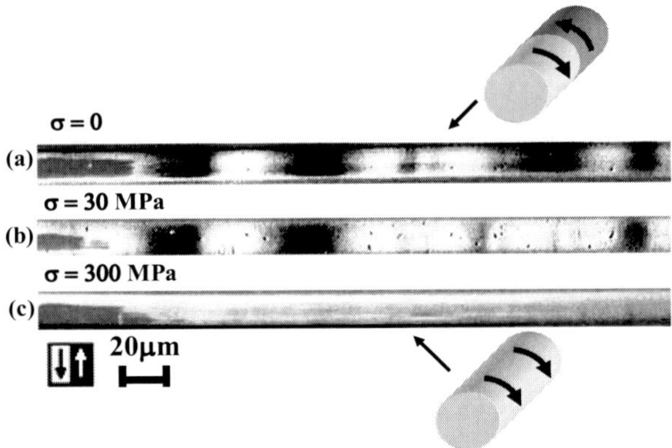

Figure 5.8   Circular domain structure in presence of external tensile stress in microwire with ρ = 0.79.

Analyzing the obtained results, we take into account the strong influence of the magnetostriction on the domain structure in the amorphous wires. The circular bistability effect is related to the formation of circular domain walls. The circular switching field should be proportional to the energy γ required to form the circular domain wall. The circular switching field is related to the magnetoelastic anisotropy as given by [7]

$$H_{SW} \propto \gamma \propto [(3/2)A_{ex}\lambda_S(\sigma_a + \sigma_r)]^{1/2}, \tag{5.6}$$

where σ_a is applied tensile stress and σ_r is the internal tensile stress. So, the switching field must be proportional to $\sigma_a^{1/2}$. The experimental HSW dependence on the tensile stress has been plotted as a function of the square root applied stress σ_a (Fig. 5.9). Good fitting of the experimental points by the linear dependence takes place for microwires with ρ of 0.94, 0.82 and 0.79 and for microwire with ρ of 0.98 for σ_a > 50 MPa. The deviation from the linear dependence observed for the microwire with ρ of 0.98 is probably related to the extremely small thickness of glass coating in this wire.

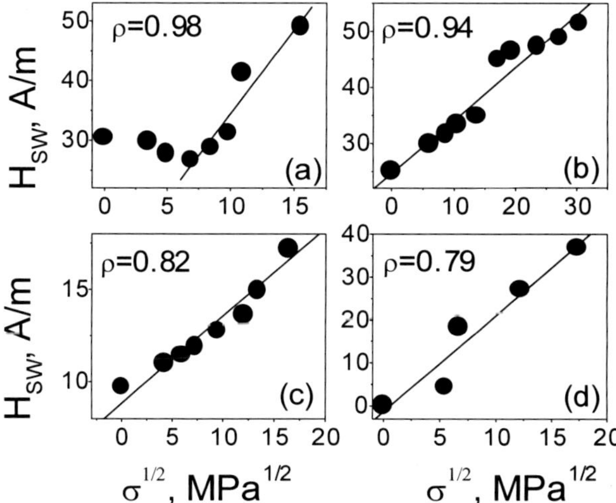

Figure 5.9 Circular switching field as a function of the square root external tensile stress.

The effect of the tensile stress on the circular switching field is associated with the strong influence of the circular magnetoelastic anisotropy on the domain structure of the microwire. It was found that the tensile stress–induced circular bistability is related to the modification of the surface domain structure from circular multi-domain configuration to mono-domain one.

References

1. Zhukova V, Chizhik A, Zhukov A, Torkunov A, Larin V, and Gonzalez J (2002), *IEEE Trans. Magn.*, **38,** 3090.

2. Chizhik A, Gonzalez J, Zhukov A, and Blanco JM (2003), *J. Phys. D: Appl. Phys.*, **36,** 419.

3. Narita K, Yamasaki J, and Fukunaga H (1980), *IEEE Trans. Magn.*, **16,** 435.

4. Ponomarev BK and Zhukov AP (1984), *Sov. Phys. Solid State*, **26,** 2974.

5. Mohri K and Takeuchi S (1982), *J. Appl. Phys.*, **53,** 8386.

6. Kraus L, Kane SN, Vazquez M, Rivero G, Fraga E, and Hernando A (1994), *J. Appl. Phys.*, **75,** 6952.

7. Gonzalez J, Murillo N, Larin V, Barandiaran JM, Vazquez M, and Hernando A (1997), *IEEE Trans. Magn.*, **33,** 2362.

8. Aragoneses P, Blanco JM, Dominguez L, Gonzalez J, Zhukov A, and Vazquez M (1998), *J. Phys. D: Appl. Phys.,* **31,** 3041.

9. Cobeno AF, Zhukov A, Arellano-Lopez AR, Elias F, Blanco JM, Larin V, and Gonzalez J (1991), *J. Mater. Res.,* **14,** 3775.

10. Blanco JM, Barbon PG, Gonzalez J, Gomez Polo C, and Vazquez M (1992), *J. Magn. Magn. Mater.,* **104–107,** 137.

11. Chizhik A, Zhukov A, Blanco JM, and Gonzalez J (2002), *J. Magn. Magn. Mater.,* **249,** 27.

12. Zhukova V, Usov NA, Zhukov A, and Gonzalez J (2002), *Phys. Rev. B,* **65,** 134407.

13. Chizhik A, Yamasaki J, Zhukov A, Gonzalez J, and Blanco JM (2004), *J. Magn. Magn. Mater.,* **272–276,** e499.

Chapter 6

Effect of High-Frequency Driving Current on Magnetization Reversal in Co-Rich Amorphous Microwires

6.1 Introduction

Amorphous glass-covered microwires attract wide science interest because of their unique magnetic properties. The giant magneto-impedance effect (GMI) [1, 2], recently discovered in these wires, became a topic of intensive research in the field of applied magnetism due to its potential application in magnetic sensors [3, 4]. The GMI is based on the change of the dynamic circular magnetization caused by the presence of a DC axial and circular magnetic field. Therefore, the investigation of the magnetization reversal in the surface area of the microwires in the presence of high-frequency (HF) electric current flowing through the microwire should be very interesting. Earlier, the remagnetization process in these materials has been studied under quasi DC condition or under the driving magnetic field of high frequency. Taking into account that the GMI effect is mainly a surface effect, investigations

Magnetic Microwires: A Magneto-Optical Study
Alexander Chizhik and Julian Gonzalez
Copyright © 2014 Pan Stanford Publishing Pte. Ltd.
ISBN 978-981-4411-25-7 (Hardcover), 978-981-4411-26-4 (eBook)
www.panstanford.com

have been performed using the magneto-optical Kerr effect (MOKE). The application of the MOKE for the study of amorphous wires and microwires demonstrated the advantages of this method for the investigation of magnetization reversal in the surface of non-plane samples [5, 6]. The purpose of this work is to study the influence of HF electric current on the process of the surface magnetization reversal in DC axial and circular magnetic field.

6.2 Experimental Details

Glass-covered amorphous microwires of nominal composition $Co_{67}Fe_{3.85}Ni_{1.45}B_{11.5}Si_{14.5}Mo_{1.7}$ (metallic nucleus radius R = 8.6 μm, glass coating thickness T = 0.6 μm), supplied by TAMAG Iberica S.L., were obtained by the Taylor–Ulitovski method [7, 8]. The experiments were performed using a transverse magneto-optical Kerr effect in the axial and circular magnetic field (Fig. 6.1). A DC circular magnetic field (H_{CIRC}) was produced by an electric current flowing through the wire. A DC axial magnetic field (H_{AX}) was produced by a pair of Helmholtz coils. During the experiments, an electric current of frequency f ranging from 4 kHz to 80 MHz and of amplitude up to 8 mA was also applied to the wire. Our experiments show that the electric current of such amplitude does not produce essential Joule heating. A polarized light from the He–Ne laser was reflected from the surface of the microwire to the detector. The beam diameter was of 0.8 mm. For the transverse Kerr effect, the intensity of the reflected light is proportional to the magnetization oriented perpendicularly to the plane of the light. In the present experiments, the circular magnetization is the magnetization that is perpendicular to the plane of the light. Therefore, the light intensity is proportional to the circular projection of the magnetization in the surface area of the wire (see Fig. 6.1).

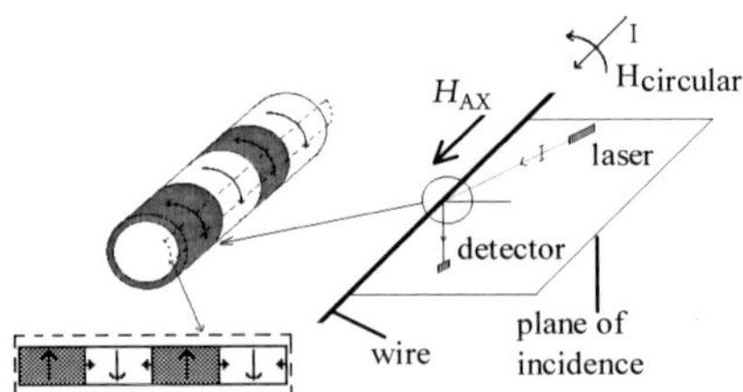

Figure 6.1 Schematic picture of experiment.

6.3 Surface Magnetization Reversal in Axial Magnetic Field

Figure 6.2 shows the influence of HF electric current (f = 4 kHz and 10 MHz) flowing along the microwire on the MOKE dependence on the axial magnetic field. The insets show how the magnetization in the surface area of the microwire changes during the magnetization reversal. The schematic pictures are presented for Figs. 6.2a,c. The magnetization reversal in the absence of HF

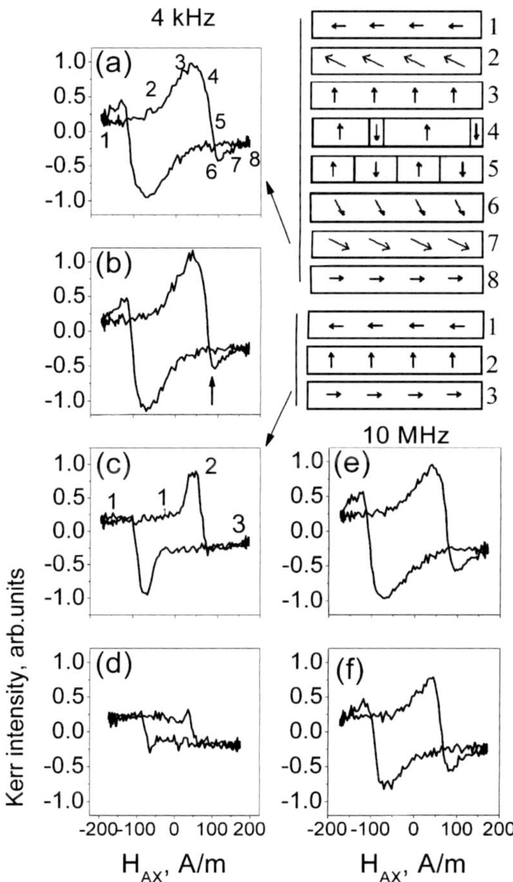

Figure 6.2 Transverse Kerr effect dependence on the axial magnetic field in the presence of HF electric current: (a–d) 4 kHz, (a) 0 mA, (b) 0.38 mA, (c) 0.76 mA, (d) 1.35 Ma, (e–f) 10 MHz, (e) 0 mA, (f) 6.7 mA.

current is demonstrated in Figs. 6.2a,e. In the first stage of the surface magnetization reversal, when the external axial magnetic field H_{AX} increases, a rotation of the magnetization from axial to circular direction in the outer shell of the wire is observed (schematic pictures 1–3, Fig. 6.2). In the second stage, the jump of the circular magnetization takes place (schematic pictures 4–6, Fig. 6.2). During this jump, the direction of the circular magnetization reverses (picture 6, Fig. 6.2). This jump is related to the circular magnetic bistability effect discovered earlier in this microwire [9]. In the third stage, the magnetization rotation from the circular to the axial direction is observed (pictures 6–8, Fig. 6.2).

The application of the HF current causes the transformation of axial hysteresis loop. The degree of this transformation depends on the value of the frequency. Two series of Figs. 6.2a,d,e,f demonstrate the HF current influence for 4 kHz and for 10 MHz. The decrease of the coercive field, H_{C-AX} and I_{MAX} takes pace. The coercive field H_{C-AX} should be considered as the field at which the drastic change of the circular magnetization starts in the presence of the axial magnetic field. I_{MAX} is the maximum value of the intensity of the Kerr signal during the magnetization reversal in axial magnetic field.

Figures 6.3 and 6.4 demonstrate the H_{C-AX} and the I_{MAX} dependencies on the HF current (I_{HF}). The decrease of the coercive field and the maximum value of the Kerr intensity is observed on all the studied interval of frequencies. For the frequency of 4 kHz, the $I_{MAX}(I_{HF})$ dependence has a small maximum at the beginning of the curves. The influence of the HF current on the coercive field becomes weaker with the increase of the frequency. We consider that $I_{MAX} \propto M_{MAX}$, where M_{MAX} is the maximum value of the circular magnetization, because as it was mentioned above, the Kerr signal is proportional to transverse magnetization. Therefore, the change of the I_{MAX} reflects the change of the M_{MAX}. The HF electric current initiates the HF oscillation of circular magnetization. When the HF current increases, the HF circular magnetization increases. Consequently, the part of the circular magnetization, which is not oscillate with the high frequency, decreases.

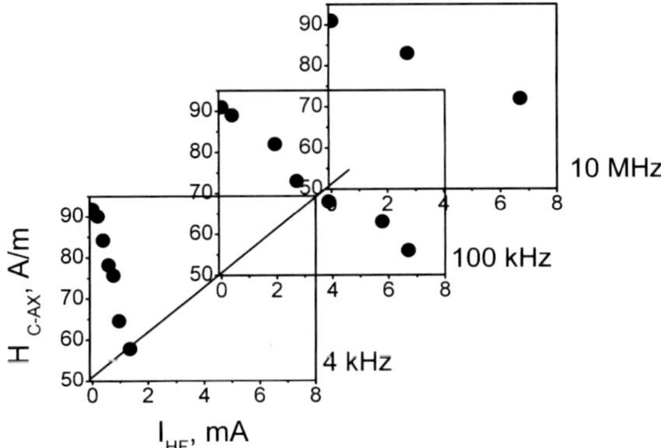

Figure 6.3 Coercive field dependence on the amplitude of HF electric current, which was obtained in axial magnetic field.

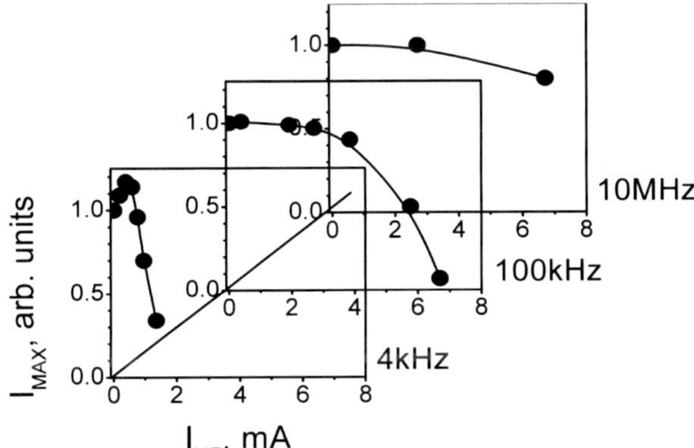

Figure 6.4 I_{MAX} dependence on the amplitude of HF electric current.

The influence of the oscillation of the circular magnetization on the nucleation of circular domains should be also taken into consideration. We suppose that the mechanism of the overcoming of the energy barrier related to the nucleation of the circular domains has a thermo-activated [11, 12] character. The oscillation of the circular magnetic field probably increases the kinetic energy of the domain wall growing the probability for the overcoming of the energy barrier related with the critical size of the reversed domain that changes the coercive field.

We considered that the magnetization reversal process in the surface area of the wire happens as the formation and the propagation of a circular domain. Therefore, the coercive field $H_{C\text{-}AX}$ is determined by the circular domain nucleation and domain walls pinning. The expression for $H_{C\text{-}AX}$ can be presented as [10]

$$H_{C\text{-}AX} = \alpha/M_{MAX} - N\, M_{MAX}, \tag{6.1}$$

where the first term is related to the losses of energy at nucleation process and the motion of the domain walls, and the second with the demagnetizing field of the nucleus (N is the demagnetizing factor). For the case of the domination of the nucleation process, $\alpha = \sigma/v^{1/3}$ (within numerical factor), where σ is the energy of the domain wall, v is the critical volume of the nucleus. For the domination of the pining of domain walls, coercive field $H_{C\text{-}AX}$ is also expressed by Eq. (6.1), in which α is a barrier that domain wall overcomes during the movement.

The $H_{C\text{-}AX} \times I_{MAX}$ dependence on I_{MAX}^2 (Fig. 6.5) has been plotted for the confirmation of our assumption relatively the peculiarities of the mechanism of the magnetization reversal. This dependence could be considered as the analogy of $H_{C\text{-}AX} \times M_{MAX}$ (M_{MAX}^2) dependence taking into account that $I_{MAX} \propto M_{MAX}$. We have obtained a good fitting of the experimental points by the linear function ($y = A + Bx$, $A = 8$, $B = 73$). This shows the strong relation between the circular magnetization and circular switching field.

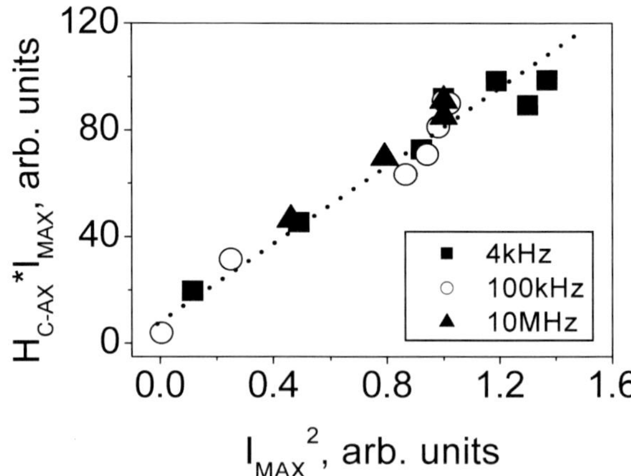

Figure 6.5 $H_C \times I_{MAX}$ dependence on I_{MAX}^2. The line is the result of the calculation.

It is necessary to note that the Kerr effect loops presented in Fig. 6.2 demonstrate the change of the mechanism of the magnetization reversal that reflects in the change of the shape of the hysteresis curve in the presence of HF current. The sharp jump (Fig. 6.2c, schematic pictures 1 and 2) replaces the monotonic change of the Kerr signal in the first stage of the magnetization reversal (Figs. 6.2a, schematic pictures 1–3). The HF circular magnetic field induces the jump of magnetization from the axial direction (Fig. 6.2c, picture 1) to the circular direction (Fig. 6.2c, picture 2) suppressing the process of the rotation of the magnetization. Because of the decrease of the influence of the HF current with the frequency increasing, the strong transformation of the hysteresis loop was not observed at the frequency of the 10 MHz (Figs. 6.2e,f) in the used range of the amplitude of the HF current. The larger amplitude of the HF current was not applied because of possible heating of the sample.

One important detail that should be indicated is the disappearance of the small peak, which is marked in Fig. 6.2b by the arrow. It is associated also with the change of the mechanism of the magnetization reversal. This peak is related with the defined mechanism of the surface magnetization reversal, as it was shown in our modeling [13]. This means that the sharp jump occurs between two circular magnetic domains (Fig. 6.2a, schematic pictures 3–6). When the HF current increases, this peak disappears (Fig. 6.2c–e) and only the jumps between the circular and axial directions (Fig. 6.2c, schematic pictures (2, 3)), but not between two circular domains, take place.

In the frame of the discussion of the disappearance of the small peak, the observed effect of the HF electric current on the surface magnetization reversal has been compared with the founded earlier circular magnetic bistability induced by the tensile stress [14]. This effect has been observed in glass-coated amorphous microwires of nominal composition $Co_{67}Fe_{3.85}Ni_{1.45}B_{11.5}Si_{14.5}Mo_{1.7}$ (metallic nucleus radius 11.2 μm, glass coating thickness 3 μm). Figure 6.6 presents the comparison of these effects (Figs. 6.6a–c show the influence of the tensile stress and Figs. 6.6d–f the influence of HF current). Two studied wires have the same composition but different thickness of glass coating. The glass coating introduces additional internal stress due to difference between the thermal expansion coefficients of glass coating and metallic nucleus. The microwires of the same composition show different magnetic properties in the absence of tensile stress and HF current because of different magnetoelastic energy [15].

As observed in Fig. 6.6, the HF current has the opposite effect on the surface magnetization reversal as compared to the tensile stress influence: the tensile stress causes the increase of circular magnetization and the HF current causes the decrease of circular magnetization. It is interesting to compare two intermediate curves presented in Figs. 6.6b,e. Being generally similar, these hysteresis loops have a difference in the magnetization reversal process (see insets of schematic domain structures in Fig. 6.6). The question is about the presence (Fig. 6.6b) or absence (Fig. 6.6e) of the above-mentioned small peak. This difference confirms that the HF electric current not only suppresses the circular magnetization but also changes the mechanism of the surface magnetization reversal.

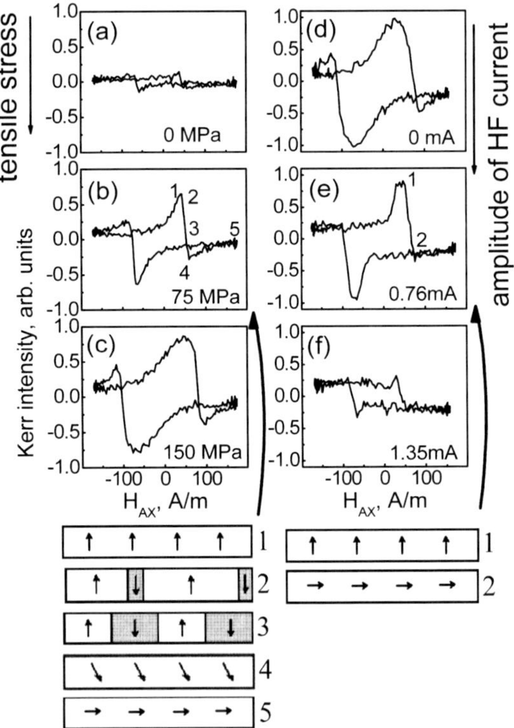

Figure 6.6 Comparison of the tensile stress (a–c) and the HF electric current (4 kHz) (d–e) influence on the surface magnetization reversal.

Also we would want to discuss the peak in I_{MAX} at small high-frequency current (4 kHz results, Fig. 6.4). We suppose that the one of the possible reason for this effect is a small inclination from the circular direction of the magnetization in the outer shell, which exists in as-quenched microwire. The small-oscillated circular magnetic field induces the rotation of equilibrium magnetization into the circular direction. This rotation results in a small growth of the peak value of the magneto-optical signal that was observed in the experiment.

6.4 Surface Magnetization Reversal in Circular Magnetic Field

Figure 6.7 presents the transverse Kerr effect dependence on the circular magnetic field H_{CIRC} with the amplitude of HF electric current (I_{HF}) as a parameter (f = 100 kHz). The H_{CIRC} value has been obtained from the value of the low-frequency current using the formula for circumferential field $H_{CIRC} = Ir/2\pi R^2$. The calculation has been performed for the surface of the metallic nucleus, i.e., for $r = R$.

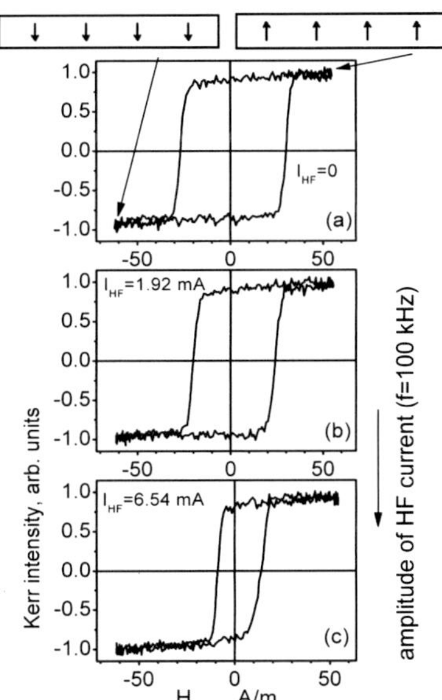

Figure 6.7   Transverse Kerr effect dependence on the circular magnetic field with the amplitude of the HF electric current (I_{HF}) as a parameter.

The values of the I_{HF} are shown in Fig. 6.7. In the absence of HF electric current (Fig. 6.7a), the shape of the hysteresis loop is perfectly rectangular with sharp vertical areas associated with quick enough reversal of circular magnetization. These reversals could be attributed to the circular magnetic bistability related to the large Barkhausen jumps between two states with opposite directions of circular magnetization (the schematic pictures of the circular domain structure are presented as inset in Fig. 6.7). The application of HF electric current causes the decrease of the circular coercive field ($H_{C\text{-}CIRC}$). The circular coercive field should be considered as the field at which the large Barkhausen jump starts.

Figure 6.8 shows the HF electric current dependencies of $H_{C\text{-}CIRC}$, obtained for different values of current frequency. The influence of the HF current on the coercive field becomes weaker as increasing the frequency of HF current and at the frequency of 80 MHz is almost negligible.

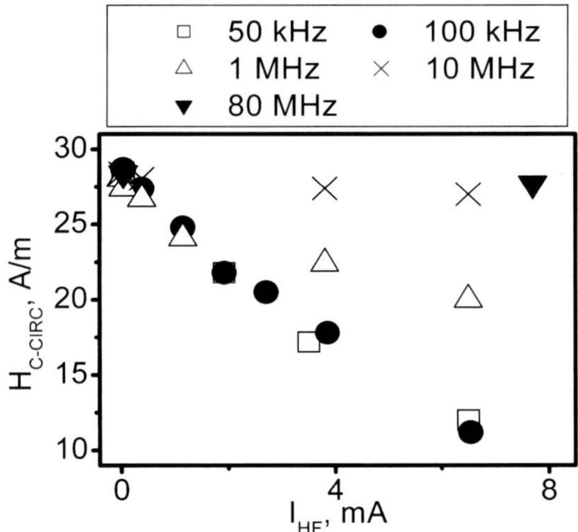

Figure 6.8   Coercive field dependence on the amplitude of HF electric current, which was obtained in circular magnetic field.

Figure 6.9 presents the effect of circular magnetic bistability induced by the HF electric current. When the HF electric current is absent, the rectangular hysteresis loop related to the jump between two states with opposite directions of circular magnetization is observed (Fig. 6.8a). Obviously, the circular bistability disappears for the circular magnetic field below the value of the circular switching field (Fig. 6.9b). The application of HF electric current causes the appearance of rectangular circular hysteresis loop (Fig. 6.9c). This effect has a threshold character.

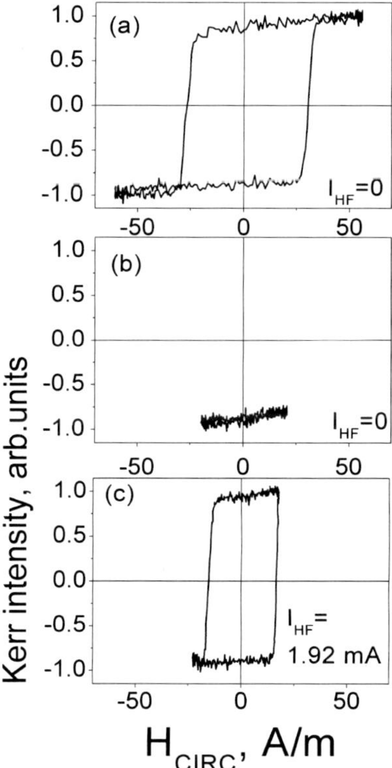

Figure 6.9 Demonstration of the HF current–induced appearance of the circular bistability.

It is worth to note that the main difference of these experiments from the experiments in axial magnetic field is that the amplitude of the magneto-optical signal does not decrease in the presence of HF current (Fig. 6.7). This means that the saturation value of the circular magnetization reversal does not change under the HF current. Contrary to the axial magnetic field, the low-frequency circular magnetic field stabilizes the surface circular magnetization and interferes its reduction in the presence of HF electric current. Therefore, we consider that the main reason for the effects observed in circular magnetic field is the influence of the HF electric current on the nucleation of the circular magnetic domains and on the motion of the domain walls, but not the reduction of the circular magnetization. We supposed that, as for the experiments in axial field, the overcoming of the energetic barrier related to the nucleation has been determined by the thermoactivation mechanism. The oscillation of the circular magnetization induced by the HF electric current brings up the probability of this overcoming and the magnetization reversal starts in the circular magnetic field below the ordinary coercive field. The amplitude of the oscillation increases with the amplitude of the HF current that raises this probability and from another side makes lower the circular coercive field (Fig. 6.7). When the circular magnetic field is not high enough to cause the jump of the circular magnetization (Fig. 6.9b), the oscillation of the magnetization makes it possible (Fig. 6.9c).

The domain wall motion in the presence of the thermoactivation mechanism is connected with the overcoming of the energetic barrier related to the domain wall pining. The HF current causes the oscillation of domain wall that raises the probability of this overcoming, and in this way also affects the circular coercivity. Since the domain wall possesses some effective mass, **the amplitude of the domain wall oscillation decreases with the frequency increase** when the amplitude of the HF current is keeping constant. Taking into account that the overcoming of the energetic barrier is related in some way with the amplitude of the domain wall oscillation, the observed decrease of the HF current influence on the coercive properties with the increasing of the frequency (Fig. 6.8) could be explained by this diminishing of the amplitude of the domain wall vibration. It is necessary to note that the threshold character of the HF current–induced appearance of circular magnetic bistability is the confirmation of the fact that the studied process of the magnetization reversal in circular direction is related to the overcoming of the energetic barrier.

It is necessary to note that the magnetization reversal in the two used experimental configurations (DC axial magnetic field and DC circular magnetic field) is basically determined by the specific domain structure of the microwires. It is known that this domain structure of the studied microwire consists of axially magnetized inner core and circularly magnetized outer shell. The complex correlation between these two parts takes place. Our previous experiments about the nucleation of the magnetic domain in the outer shell of the wire [16] have demonstrated that this process occurs in absolutely different ways in axial and circular magnetic fields.

We consider that when we apply the DC axial magnetic field, the magnetization reversal stars in the inner core that initiates the magnetization reversal in the outer shell. In the same time, the DC circular magnetic field acts directly on the circularly magnetized outer shell. In this situation, for the experimental configuration DC circular field + HF circular field the HF circular field influences directly on the nucleation of surface circular domains. For the experimental configuration DC axial field + HF circular field, in the first stage, the HF circular field influences on the reversal magnetization of the inner core and the magnetization reversal in the outer shell could be considered as the continuation of this first process.

The magneto-optical Kerr effect investigations of the surface coercive properties have been performed in Co-rich, amorphous microwires under applied HF electric current. The transformation of the axial and circular surface hysteresis loops has been found in the presence of the HF current.

In axial magnetic field, a change of the mechanism of the surface magnetization reversal as a result of the effect of the HF current has been observed: when the amplitude of the HF current is quite small, the surface magnetization reversal consists of the rotation of the magnetization from the axial to the circular direction followed by a jump between two states with circular magnetization; when the amplitude of the HF current is high enough, the magnetization reversal in the outer shell occurs as a jump between the axial and circular directions. The HF electric current causes a decrease of the coercive field in axial magnetic field and a decrease of the surface circular magnetization. A strong correlation of the coercive field and the surface circular magnetization has been obtained on all the studied interval of the frequencies. The analysis of the experimental results has been performed allowing for the nucleation and propagation of the circular magnetic domain.

In the circular magnetic field, the effect of the HF current–induced decrease of the circular coercive field has been also found, but the decrease of the surface circular saturation magnetization has not been observed. The HF current–induced appearance of circular bistability has been also found. The main reason for the effects observed in circular magnetic field is the influence of the HF electric current on the nucleation of circular magnetic domains and on the motion of domain walls, but not the reduction of the circular magnetization. We consider that the HF current increases the kinetic energy of domain wall bringing up the probability of thermoactivated overcoming of the energetic barrier related to the circular domain nucleation and the domain wall pining. In this way, the HF current reduces the circular coercivity or induces the circular bistability in small magnetic field.

References

1. Beach RS and Berkowitz AE (1994), *Appl. Phys. Lett.*, **64**, 3652.

2. Panina LV and Mohri K (1994), *Appl. Phys. Lett.*, **65**, 1189.

3. Zhukov A, Gonzalez J, Blanco JM, Vázquez M, and Larin V (2000), *J. Mater. Res.*, **15**, 2107.

4. Zhukov A, Gonzalez J, Vázquez M, Larin V, and Torcunov A (2004), *Encyclopedia of Nanoscience and Nanotechnology*, **10,** (American Scientific Publishers Valencia, California, USA).

5. Chizhik A, Gonzalez J, Zhukov A, and Blanco JM (2002), *J. Appl. Phys.*, **91**, 537.

6. Raposo VJ, Gallego JM, and Vazquez M (2002), *J. Magn. Magn. Mater.*, **242–245**, 1435.

7. Taylor GF (1924), *Phys. Rev.*, **24**, 6555.

8. Vazquez M and Zhukov A (1996), *J. Magn. Magn. Mater.*, **160**, 223.

9. Chizhik A, Gonzalez J, Zhukov A, and Blanco JM (2003), *J. Phys. D: Appl. Phys.*, **36**, 419.

10. Givord D, Tanaud P, and Viadieu T (1998), *IEEE Trans. Magn.*, **24**, 1921.

11. Zhukova V, Zhukov A, Blanco JM, Gonzalez J, and Ponomarev BK (2002), *J. Magn. Magn. Mater.*, **249**, 131.

12. Cobeño AF, Zhukov A, Blanco JM, and Gonzalez J (2001), *J. Magn. Magn. Mater.*, **234**, L359.

13. Chizhik A, Zhukov A, Blanco JM, and Gonzalez J (2004), *Phys. B*, **343**, 374.

14. Chizhik A, Gonzalez J, Zhukov A, and Blanco JM (2003), *Appl. Phys. Lett.*, **82**, 610.

15. Chizhik A, Yamasaki J, Zhukov A, Gonzalez J, and Blanco JM (2004) *J. Magn. Magn. Mater.*, **272–276S**, E499.

16. Chizhik A, Gonzalez J, Yamasaki J, Zhukov A, and Blanco JM (2004) *J. Appl. Phys.*, **95**, 2933.

Chapter 7

Relation between Surface Magnetization Reversal and Magnetoimpedance

7.1 Introduction

The GMI effect is basically the surface effect, and our previous investigation using the magneto-optical Kerr effect showed that the transformation of the surface magnetic structure has a great influence on the value of the GMI ratio [1]. A special role in the correlation between the results of the magnetoimpedance and magneto-optical experiments plays the surface helical anisotropy [2]. Taking into account that the surface anisotropy is very sensitive to different types of the mechanical stresses, we have chosen a series of glass-covered microwires with different thickness of glass covering. It is known that the glass covering produces strong internal stress in the wire that leads to a transformation of the magnetic structure of the surface.

In this chapter, we establish the correlation between the surface magnetization reversal and the value of the GMI ratio in

Magnetic Microwires: A Magneto-Optical Study
Alexander Chizhik and Julian Gonzalez
Copyright © 2014 Pan Stanford Publishing Pte. Ltd.
ISBN 978-981-4411-25-7 (Hardcover), 978-981-4411-26-4 (eBook)
www.panstanford.com

glass-covered microwires. Therefore, choosing the present series of the microwires, we hope to obtain a wide variation of the shape of the surface hysteresis loops and, in turn, a variation of the value of the angle between the helical anisotropy axis and the wire axis.

7.2 Optimization of Giant Magnetoimpedance in Co-Rich Amorphous Microwires

Soft magnetic glass-coated $Co_{67}Fe_{3.85}Ni_{1.45}B_{11.5}Si_{14.5}Mo_{1.7}$ microwires with different geometric ratio, ρ, of metallic nucleus diameter to total microwire diameter $0.69 \leq \rho \leq 0.98$ and low magnetic anisotropy field (50–200 A/m) have been studied. Conventional and surface hysteresis loops have been measured by the fluxmetric and Kerr effect methods, respectively.

A maximum DC longitudinal magnetic field, H_{MAX}, up to 2400 A/m was supplied by a long solenoid. The dependence of the magnetoimpedance ratio, $\Delta Z/Z$, on axial magnetic field, H, has been investigated for the frequency, f, range 0.06–15 MHz and driving current amplitude, I, of 0.75–2 mA. Bulk hysteresis loops of three microwires are shown in Fig. 7.1. As can be observed from this figure, the magnetic anisotropy field, H_k increases with decreasing ratio ρ, i.e., with the increase of the glass coating thickness.

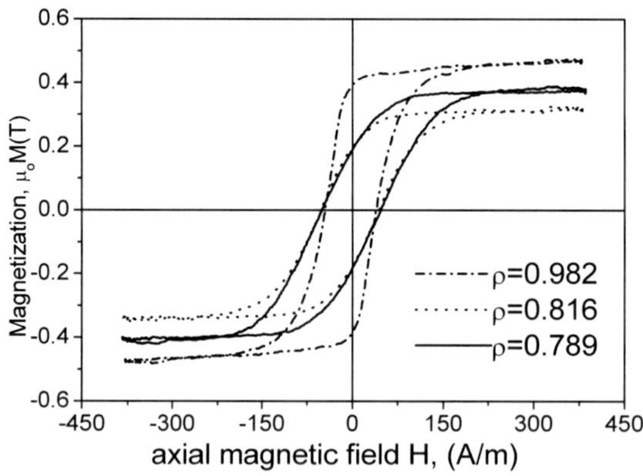

Figure 7.1 Bulk hysteresis loop of three samples with ρ as a parameter.

The $(\Delta Z/Z)(H)$ dependencies measured at f = 10 MHz and I = 0.75 mA are presented in Fig. 7.2. A maximum relative change in the GMI ratio, $\Delta Z/Z$, up to around 615% is observed at $f = 10$ MHz and $I = 0.75$ mA in the sample with $\rho = 0.98$.

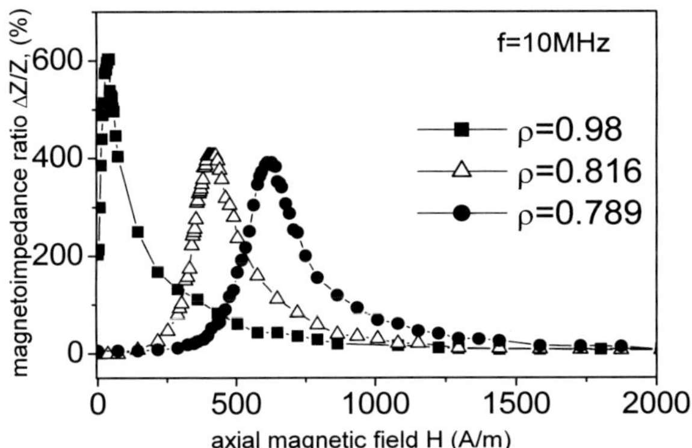

Figure 7.2 Axial field dependence of $\Delta Z/Z$ at $f = 10$ MHz and $I = 0.75$ mA in microwire with ρ as a parameter.

As shown in Fig. 7.2, the field corresponding to the maximum of the GMI ratio, H_m increases and $(\Delta Z/Z)_m$ decreases with ρ. Such $H_m(\rho)$ dependence should be attributed to the effect of internal stresses, σ, on the magnetic anisotropy field. Indeed, the value of the DC axial field corresponding to the maximum of the GMI ratio, H_m, should be attributed to the static circular anisotropy field, H_k [3, 4]. The estimated values of the internal stresses in these amorphous microwires are of the order of 1000 MPa, depending strongly on the thickness of glass coating and metallic nucleus radius [5]. Such elevated internal stresses give rise to a drastic change of the magnetoelastic energy, $K_{me} \approx 3/2\lambda_s\sigma_i$, even for small changes of the glass coating thickness at fixed metallic nucleus diameter. Consequently, such change of the ratio ρ should be related to the change of the magnetostriction constant [4]:

$$\lambda_s = (\mu_o M_S/3)(dH_k/d\sigma), \tag{7.1}$$

where M_S is the saturation magnetization.

It has been demonstrated that the spatial magnetization distribution close to the surface is very sensitive to the internal or applied stresses [6, 7].

It was shown that the $(\Delta Z/Z)(H)$ dependence is mainly determined by the type of magnetic anisotropy [7]. Circumferential anisotropy leads to the observation of the maximum of the real component of wire impedance (and consequently of the GMI ratio) as a function of the external magnetic field. In the case of axial magnetic anisotropy, the maximum value of the GMI ratio corresponds to zero magnetic field [7], i.e., results in monotonic decay of GMI ratio with H. Therefore, the important contribution of the non-diagonal components of the permeability tensor is expected for the samples with well-defined maximum in the axial field dependence of the GMI ratio [7].

The MOKE hysteresis loop reflects the axial field dependence of circular magnetization (see Fig. 7.3) in the outer shell of the wire.

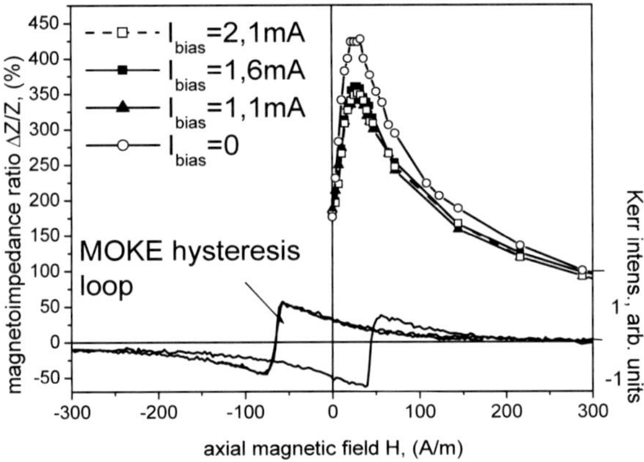

Figure 7.3 $\Delta Z/Z$ and MOKE hysteresis loop of microwire with $\rho = 0.98$.

Observed MOKE hysteresis loops can be interpreted in the following way: the absence of the circular magnetization under axial magnetic field above 150 A/m reflects the axial alignment of the magnetization in the surface layer at this magnetic field range. The monotonic increase of the magnetization with decreasing the field below 150 A/m is related to the magnetization rotation from the axial to the circular direction. The relatively sharp change of magnetization (at around ±50 A/m) could be attributed to the nucleation of new domains with the opposite circular magnetization (appearance of bamboo-like domain structure), and growth of these new domains through the domain walls propagation, until the single circular domain structure with the opposite circular magnetization is appeared. Further increase of H results in the magnetization rotation toward the axial direction.

A correlation has been observed between the $(\Delta Z/Z)(H)$ and the surface axial hysteresis loops: a maximum of GMI ratio occurs approximately at the same axial magnetic field as the sharp change of magnetization on the MOKE loop (see Fig. 7.3).

7.3 Circular Surface Magnetization Reversal and Magnetoimpedance

In this chapter, four microwires with different geometric ratios, ρ, 0.79, 0.88, 0.90, and 0.93, have been studied (nominal composition $Co_{69.5}Fe_{3.9}Ni_1B_{11.8}Si_{10.8}Mo_2$). The experiments have been performed using transverse MOKE. The electrical impedance of the microwires was evaluated by means of a network analyzer.

Figure 7.4 presents the MOKE dependencies on the AC electric current flowing along the wires with a DC axial field as a parameter. Figure 7.2 presents the MOKE effect dependencies on the AC axial field.

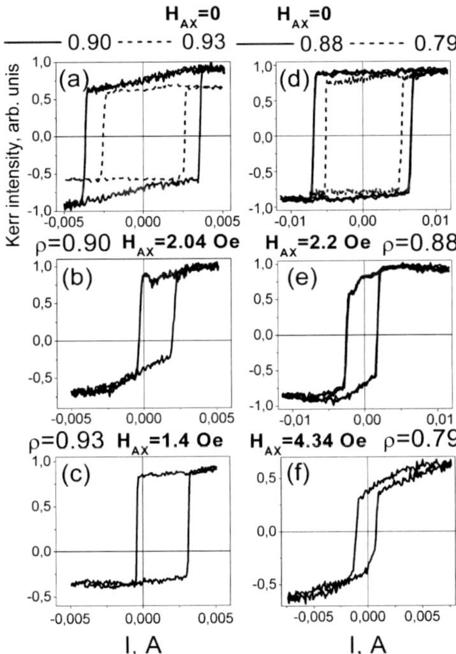

Figure 7.4 Transverse Kerr effect dependencies on electric current (circular magnetic field) with the axial bias field as a parameter.

When the DC axial magnetic field is absent, the shape of the circular hysteresis loop is perfectly rectangular (Figs. 7.4a,d). The DC axial magnetic field initiates the transformation of the circular hysteresis loop in two different ways.

Figures 7.4b,c demonstrate the shape of the hysteresis loop for the wire with ρ = 0.90 and 0.93 in the presence of the axial DC bias magnetic field. The application of the DC axial magnetic field causes the asymmetrical change of the switching field, H_{SW} (associated with the switching current). The value of H_{sw}^{+} increases when the value of H_{sw}^{-} decreases. This effect is realized in the observed "shift" of the hysteresis loop along the X axis. When the sign of the DC axial magnetic field changes to the opposite one, the picture reverses: H_{sw}^{+} decreases and H_{sw}^{-} increases.

The transformation of the circular hysteresis loop for the wires with ρ = 0.79 and 0.88 occurs in another way. The absolute values of H_{sw}^{-} and H_{sw}^{+} decreases symmetrically and the "shift" effect is not observed (Figs. 7.4e,f)). The application of the negative axial magnetic field causes the transformation of the hysteresis loop in the same way as the positive one.

The difference also takes place in the hysteresis loops obtained in the AC axial magnetic field (Fig. 7.5). The surface magnetization reversal in Co-rich microwires consists of rotation of the magnetization and the nucleation of the circular domains [9]. First, when the external axial magnetic field increases the rotation of the magnetization from axial direction to circular direction is observed. After that, the sharp jump of the signal with the change of the signal sign takes place. This is related to the nucleation of circular domain. For the wires with ρ = 0.90 and 0.93, the jump of magnetization in high enough. For the wires with ρ = 0.79 and 0.88, the magnetization reversal consists mainly of the rotation of the magnetization. The jump of the magnetization related to the nucleation of the circular domains is relatively small.

The axial field dependence of the magnetoimpedance (MI) ratio $\Delta Z/Z$ measured at fixed AC driving current amplitude of 1 mA and frequency 10 MHz is presented in Fig. 7.6. The MI ratio has been defined as $\Delta Z/Z = [Z(H) - Z(H_{max})]/Z(H_{max})$. The maximum value of the MI ratio has been obtained for the wire with ρ = 0.90.

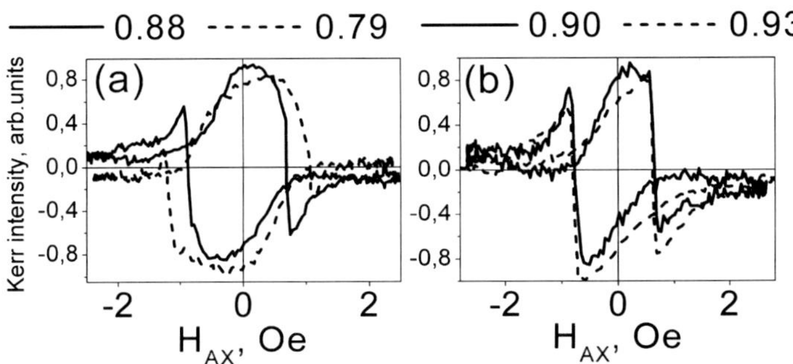

Figure 7.5 Transverse Kerr effect dependencies on axial magnetic field for four wires with different thickness of glass covering.

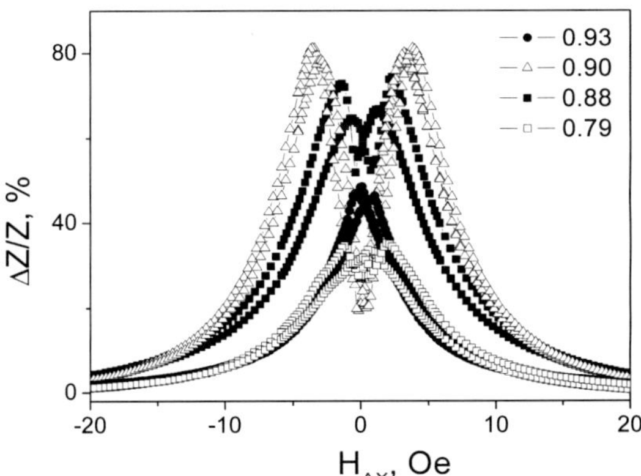

Figure 7.6 Axial field dependence of the MI for four wires with different thickness of glass covering.

The main difference between the hysteresis loops presented in Fig. 7.4 is the existence of the shift of hysteresis loop for the $\rho = 0.9$ and 0.93 and the absence of this shift for the $\rho = 0.79$ and 0.88. For the wires with $\rho = 0.79$ and 0.88 the DC axial magnetic field causes only the inclination of the magnetization in the outer shell of the wire toward the axial direction (Figs. 7.4e,f). This inclination takes place in the same way in circular domains (+) and (−). The H_{sw}^- and H_{sw}^+ values decrease in the presence of axial DC field because the DC magnetic field favors the nucleation of a circular domain of two types.

For the wires with $\rho = 0.90$ and 0.93, the DC magnetic field favors the nucleation of a circular domain of one type and delays the nucleation of the domains of the other type. The observed correlation of the signs of the axial magnetic field and the nucleated circular domain is the main difference of these wires. The axial field has direct influence on the nucleation of the axial domain. The observed shift of the surface loop reflects the strong correlation between the axial domain of defined direction and the circular domain of the defined direction in the outer shell.

The main conclusion from this experiment is that for the wires with ρ = 0.79 and 0.88, the DC axial field initiates mainly the inclination of the magnetization in the outer shell. For the wires with ρ = 0.90 and 0.93, the DC axial magnetic field supports the nucleation of circular domain but not the rotation of the magnetization. This process of nucleation is accompanied by the formation and the motion of circular domain walls that leads to the increase of the circular permeability.

In AC axial field (Fig. 7.5), for the wires with ρ = 0.90 and 0.93, the nucleation of circular domain occurs just after the magnetization has reached the circular direction. For the wires with ρ = 0.79 and 0.88, the rotation of magnetization continues after the reaching of the maximal point of the circular magnetization and the nucleation occurs later. In this situation, the magnetization reversal is determined basically by the rotation of the magnetization and the value of the jump of magnetization related to the nucleation of the circular domain is small. From this experiment, we can conclude that the contribution of the domain walls motion to the magnetization reversal is smaller for the wires with ρ = 0.79 and 0.88 than for the wires with ρ = 0.90 and 0.93.

Following is the explanation of the above-mentioned MOKE experiments. The jump of circular magnetization and the circular domain wall dynamics related to it, determines essentially the circumferential permeability [8, 13]. In turn, MI effect is closely associated with the circumferential permeability $\mu_{circ} = dM_{circ}/dH$. In the MOKE experiments the jump of the circular magnetization is smaller for the wires with ρ = 0.79 and 0.88 than for the wires with ρ = 0.90 and 0.93. It finds the reflection in the smaller value of the circumferential permeability and the smaller value the MI ratio.

The series of the glass-covered microwires was considered as a model one having some distribution of the GMI value. Generally, the surface magnetic structure in the glass-covered microwires originates from the magnetoelastic anisotropy associated with the internal stresses. One of the sources of these stresses is the difference of the thermal expansion coefficients of the glass coating and metallic nucleus [10].

The surface magnetization reversal has been studied in Co-rich glass-covered microwires with different thickness of glass covering, which demonstrate great difference in the MI ratio. The difference in the mechanism of the surface magnetization reversal has been discovered. The DC axial magnetic field shift of circular hysteresis loop could be considered as a characteristic attribute of high value of magnetoimpedance. The value of the jump of the circular magnetization has the correlation with the circular permeability and in such a way with the value of GMI. Therefore, we can conclude that the value of the MI could be predicted based on the analysis of the shape of the surface magnetization reversal curves.

7.4 Axial Surface Magnetization Reversal and Magnetoimpedance

An example of the axial field dependence of the magnetoimpedance measured at 10 MHz with AC current of 1 mA amplitude is presented in Fig. 7.7a for ρ = 0.93. Figure 7.7b presents the distribution of the maximal value of $\Delta Z/Z$ for the series of the studied wires. The maximal value of 80% has been obtained for the wire with ρ = 0.90.

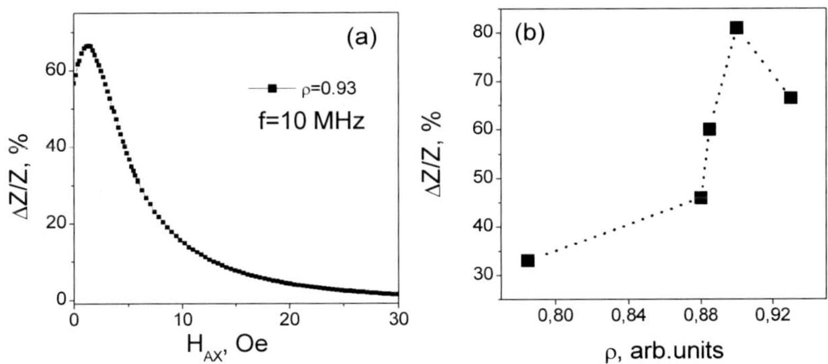

Figure 7.7 (a) Axial field dependence of the MI for the wire with for ρ = 0.93. (b) Dependence of maximal value of MI on ratio ρ.

Kerr effect hysteresis loops obtained in the AC axial magnetic field are presented in Figs. 7.8a,e. We put the attention on the transformation of the shape of hysteresis curve in the dependence of the ratio ρ.

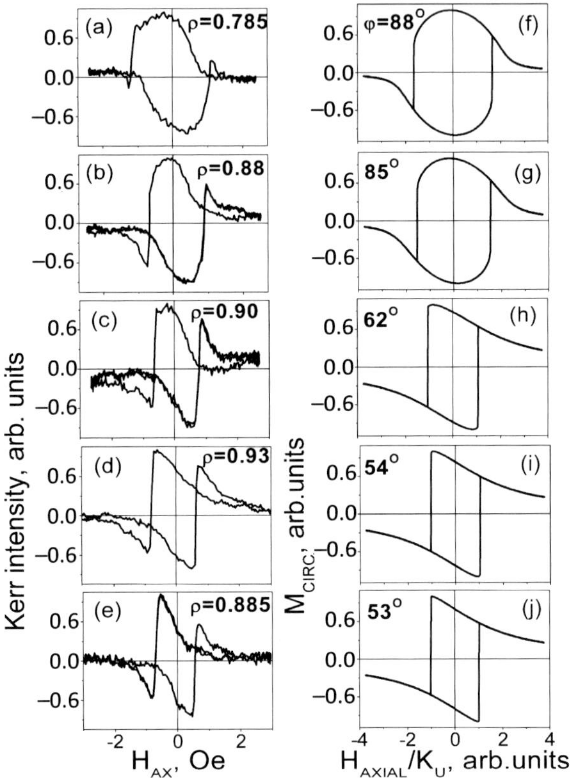

Figure 7.8 (a–e) Experimental transverse Kerr effect dependencies on axial magnetic field for series of wires with different thickness of glass covering. (f–j) Calculated dependencies of circular magnetization on normalized axial magnetic field for different angles of anisotropy direction φ.

Figures 7.8a,e are presented in the order of a decreasing contribution of the rotation to the magnetization reversal and the increase of the jump of the magnetization. For the wire with $\rho = 0.785$, the magnetization reversal consists mainly of the rotation of the magnetization and the jump of magnetization is small. For the wire with $\rho = 0.90$ and 0.93, the jump is large.

We consider that the observed transformation of the surface hysteresis loops is related to the transformation of the surface helical magnetic structure. Therefore, the calculation of the hysteresis loops has been performed taking into account an existence of a helical magnetic anisotropy in the surface area of the wire. We treat the wire surface as a two-dimensional system in our calculations, because the curvature of the area of wire surface, from which the light is reflected to detector, is about 1°, as it was mentioned above. The applied magnetic field can be expressed as a superposition of two mutually perpendicular fields (H_{AX} and H_{CIRC}). The direction of the anisotropy was changed from axial to circular direction. The expression of the energy of the system has the following form:

$$U = -K_U \cos^2(\theta - \varphi) - h \times M_S$$
$$= -K_U \cos^2(\theta - \varphi) - H_{AX} \cos(\theta) - H_{CIRC} \sin(\theta), \tag{7.2}$$

where K_U is uniaxial anisotropy constant, M_S is the saturation magnetization, φ is an angle between the anisotropy axis and the wire axis and θ is the angle between the magnetic moment and the wire axis.

The numerical calculation was done by the coherent rotation approach [11]. The hysteresis loops was computed by the minimalization of the energy term described by Eq. 7.2. For given values of the angle φ the minimalization procedure can be outlined in the following way: each time the value of the axial field was changed, the search of the value of the angle, θ, that gave the minimal value of the energy term (2) was done. The circular field component remained equal to zero all the time.

The results of the numerical analysis of Eq. 7.2 are presented in Figs. 7.8f,j and Fig. 7.9. During the comparative analysis of the results of the experiment and the calculation, we pay attention to the value and the position of the jump of circular magnetization. The GMI effect is closely associated with the circumferential permeability. Therefore, the value of the jump of circular magnetization ΔM is very important parameter for GMI.

Figure 7.9 presents the dependence of the jump of the circular magnetization ΔM on the angle of helical anisotropy and some examples of the calculated dependencies of the circular magnetization on the axial magnetic field. This dependence has a maximum for the angle of 62°. The maximum is not sharp, but for the angles close to 0° and 90°, the change of the ΔM is strong enough. We can mark the great difference in the shape of the calculated hysteresis curves: the smooth shape of the peak for the angles close to 90° means that the magnetization reversal is determined mainly by a fluent rotation of the magnetization and this rotation continues after the moment when the magnetization reaches the exact circular direction. For the angles close to 60° the jump of the magnetization occurs at the moment when the direction of the magnetization is close to the circular magnetization. Because of that the value of the ΔM is large for these angles. When the angle of helicality approximates to 0° the inclination of the magnetization from the axial direction and consequently, the value of the ΔM decreases.

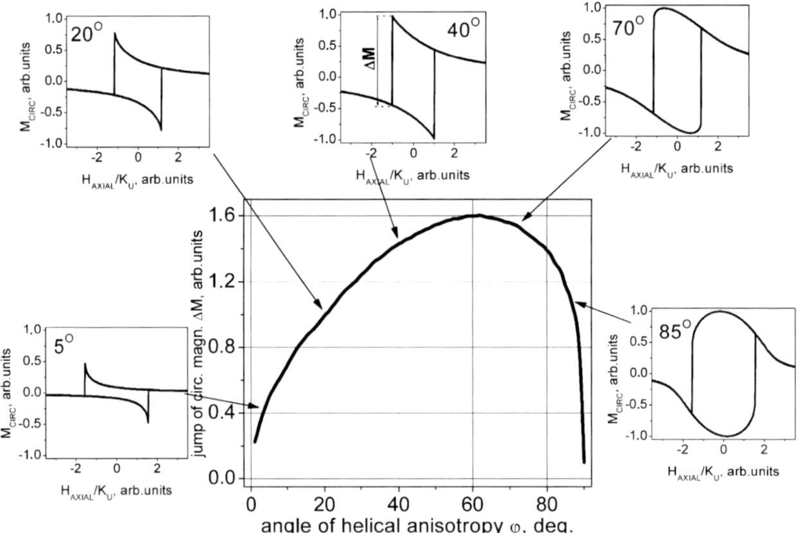

Figure 7.9 Calculated dependence of the jump of the circular magnetization ΔM on the angle of helical anisotropy φ. Insets show the calculated hysteresis loops for different value of φ.

The angles of helical anisotropy were assigned to experimental loops taking into account the value and the position of the jump of the circular magnetization ΔM (see inset to Fig. 7.9 for 40°) [12]. These parameters are very sensitive to the value of helical anisotropy. Also, in our analysis, we used the parameter $K = H_{SW}/H^*$, where H^* is the axial magnetic field at which the rotation of the magnetization starts. For example, for the wire with $\rho = 0.93$, we get K = 0.24, and for the wire with ρ = 0.785, the result is K = 0.8. The value of the parameter K reflects the relation between the processes of the rotation of magnetization and the jump of circular magnetization.

In such a way, it was determined that the calculated curves for φ = 88°, 85°, 62°, 54°, and 53° (Figs. 7.8f,j) correspond to the experimental curves with ρ = 0.785, 0.88, 0.90, 0.93, and 0.885. Having this correspondence of the experimental and calculated curves, we have prepared Fig. 7.10, in which the experimental value of GMI is presented as a function of the calculated value of the helical anisotropy. The values of ρ are also marked in this figure. The value of the GMI has a maximum of 80% for the wire with helical angle of 62° and decreases when the φ angle inclines from this amount. It is necessary to note that the observed distribution of the surface magnetic properties originates from the magnetoelastic anisotropy related to the stress induced by the glass covering [6].

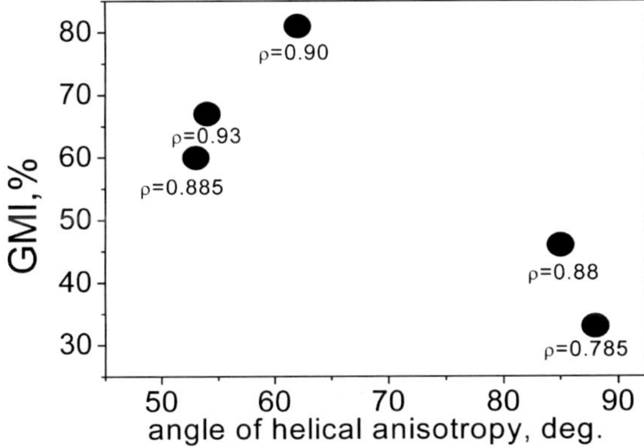

Figure 7.10 Dependence of the experimental value of GMI ratio for wires with different ρ on the calculated value of angle of anisotropy direction φ.

Generally, the GMI effect is interpreted in terms of the classical skin effect in a magnetic conductor assuming scalar character for the magnetic permeability, as a consequence of the change in the penetration depth of the AC current caused by the DC applied magnetic field. The electrical impedance, Z, of a magnetic conductor in this case is given as follows [3]:

$$Z = R_{\mathrm{DC}} kr J_0(kr)/2J_1(kr), \tag{7.3}$$

with $k = (1 + j)/\delta$, where J_0 and J_1 are the Bessel functions, r the wire's radius and δ the penetration depth given by $\delta = (\pi \sigma \mu_{\mathrm{circ}} f)^{-1/2}$, where σ is the electrical conductivity, f the frequency of the current along the sample, and μ_{circ} the circular magnetic permeability. After the analysis of the Kerr effect experiments and the calculations, we have the direct correlation between the value of ΔM and the angle of helical anisotropy δ. Now, taking into account the above-mentioned role of the jump of the circular magnetization ΔM for the GMI effect, we can establish for the first time, the correlation between the experimental value of the GMI effect and the value of the angle of helical anisotropy φ in the surface area of the wire.

To verify the above-mentioned consideration, we have performed the estimation of the MI ratio for two wires with different directions of the helical anisotropy ($\rho = 0.93$ and $\rho = 0.785$), using the formula (3). The experimental hysteresis curves presented in Figs. 7.8a,d have been differentiated. Taking into account that the intensity I obtained in the magneto-optical experiments is proportional to the circular magnetization, the results of the differentiation presented in Fig. 7.11a could be considered as the analogy of the field dependence of the circumferential permeability μ_{circ}. These two dependencies were approximated by the polynomial function and used to estimate the MI ratio (Eq. 7.3). The results of the calculation of the MI ratio are presented in Fig. 7.11b. The performed calculation demonstrates that that the maximal value of the MI ratio is higher for the microwire with helical anisotropy inclined from the circular direction ($\varphi = 54°, \rho = 0.93$) than for the microwire with helical anisotropy closed to the circular direction ($\varphi = 88°, \rho = 0.785$) that is in agreement with the results of the MI experiments (Fig. 7.7).

Comparative analysis of magneto-electric and magneto-optic experiments and the results of the calculation shows the strong correlation between the value of the GMI ratio and the shape of the Kerr effect hysteresis loops. The three parameters—the value of the GMI ratio, the value of the jump of circular magnetization, and the value of the angle of helical anisotropy—have been connected to one logical chain. It was found that the maximum value of the GMI could be achieved for the angles of helical anisotropy close to 60°.

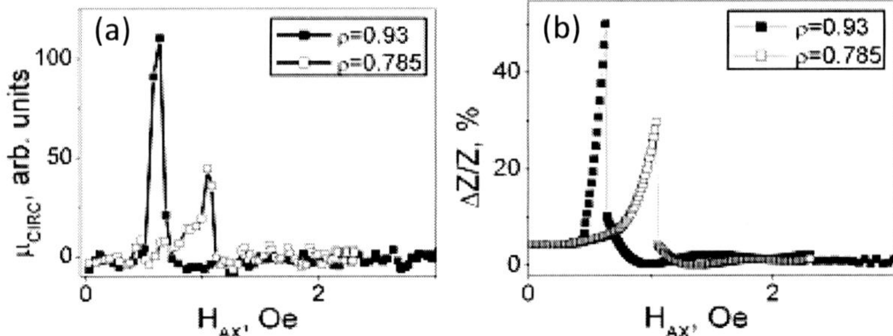

Figure 7.11 (a) Axial magnetic field dependencies of the circumferential permeability μ_{circ} extracted from the experimental Kerr effect dependencies on axial magnetic field presented in the Figs. 7.8a,d; (b) Calculated axial field dependence of the MI for two wires with different geometric ratio ρ (different angles of helical anisotropy).

7.5 Correlation of Surface Domain Structure and Magnetoimpedance

Generally, the industrial sensors utilize amorphous microwires as sensing element and are realized on so-called off-diagonal impedance, which exhibits higher sensitivity and linearity as compared with the longitudinal diagonal impedance of the conductor. The off-diagonal impedance is observed in magnetically soft wires with circumferential or helical magnetic anisotropy and is related to the cross-magnetization process m_z/h_φ [14, 15]. If the static magnetization is helical (as a result of application of external longitudinal magnetic field, for example), then the precision of the circular magnetic field h_φ created by the AC current i gives rise to appearance of the nonzero axial magnetization m_z which, in accordance with Faraday's law, induces a voltage in the pick-up coil wounded on the wire. In this way, the off-diagonal impedance can be detected. The off-diagonal impedance, besides the higher sensitivity to magnetic field comparing with the longitudinal impedance, is also more sensitive to the particularities of surface domain structure (SDS).

For example, if the SDS divides into domains with opposite circular magnetization direction (bamboo-like SDS), the voltage response averages over domains and the observed off-diagonal MI effect is very small and irregular, while the longitudinal MI effect in the same sample, that we experimentally demonstrate further, is rather high. A detailed understanding of the influence of SDS on MI properties is essential for further progress in the development of stress sensitive composite materials and sensors. In this chapter, we investigate the correlation between the SDS, which was modified by the application of stresses, and high frequency MI in amorphous microwires with circumferential and helical anisotropy. We propose a rather simple method for the determination of SDS from MI measurements.

Figure 7.12 shows the SDS of the sample in unstressed state (a) and with applied tensile stress (b). One can see that the SDS is a bamboo-like in the unstressed wire that transforms to a mono-domain after applying tensile stress. Figure 7.13 shows the SDS of the twisted wire that, as it can be seen, resulted in the appearance of mono-domain state with helical anisotropy. The anisotropy angle can be evaluated from the image in zero field (Fig. 7.13a) as the image contrast is determined by the magnetization direction. Application of the axial magnetic field H_E leads to rotation of the magnetization as shown in Figs. 7.13b,c.

We have experimentally investigated the impedance dependencies in three pieces of the same microwire to which different stresses were applied in the same way as in the magneto-optical experiment (see Table 7.1).

Table 7.1   Description of the measurements

#	Applied stress	Domain structure	Anistropy
1	No applied stresses	Bamboo	Circumferential
2	Tensile	Mono-domain	Circumferential
3	Torsional	Mono-domain	Helical

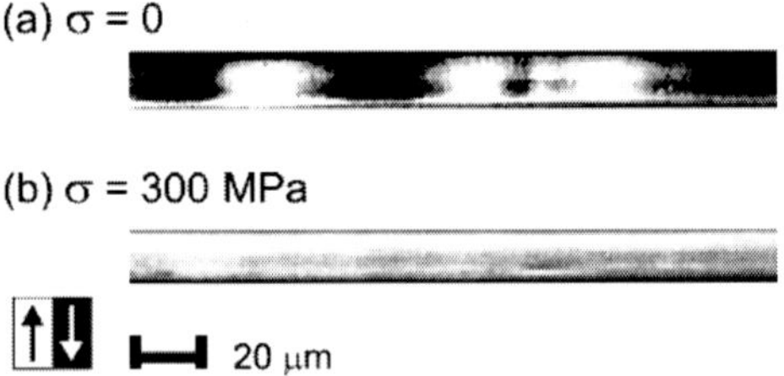

Figure 7.12   MOPM photograph of the surface domain structure of microwire in unstressed state (a) and with applied tensile stress *r* (b).

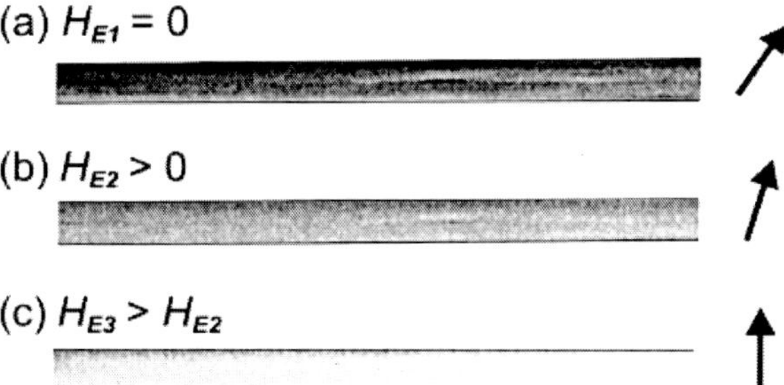

Figure 7.13   MOPM photograph of the surface domain structure obtained of microwire subjected to torsional stress. The arrows show the magnetization direction obtained from the image contrast.

The longitudinal Z_{zz} and off-diagonal $Z_{\varphi z}$ impedance components dependencies on external axial magnetic field H_E were measured in 6 mm-long pieces of amorphous glass-coated microwire. The microwires were placed in a specially designed microstrip cell. One wire end was connected to the inner conductor of a coaxial line through a matched micro-strip line while the other was connected to the ground plane.

The impedance components Z_{zz} and $Z_{\varphi z}$ were measured simultaneously using vector network analyzer in the frequency range 10–300 MHz. The longitudinal impedance of the sample $Z_w = Z_{zz}l$, where l is the wire length, was obtained from reflection coefficient S_{11} and the off-diagonal impedance $Z_{\varphi z}$ was measured as transmission coefficient S_{21} as a voltage induced in a 2 mm-long pick-up coil wounded over the wire. The static bias field H_B was created by the DC current I_B applied to sample through the bias-tee element.

The other experimental details are given in Ref. 15. The measurements of the real parts of longitudinal Z_{zz} and off-diagonal $Z_{\varphi z}$ impedances at frequency of 30 MHz are shown in Fig. 7.14. Similar dependencies were observed in the whole frequency range. However, at higher frequencies, we observed the increase of Z_{zz} component. The $Z_{\varphi z}$ impedance, on the contrary, decreases that is probably related to the reactance of the pick-up coil. The graphs in Fig. 7.14 show both ascending and descending branches of the field dependencies so that the magnetic hysteresis can be evaluated.

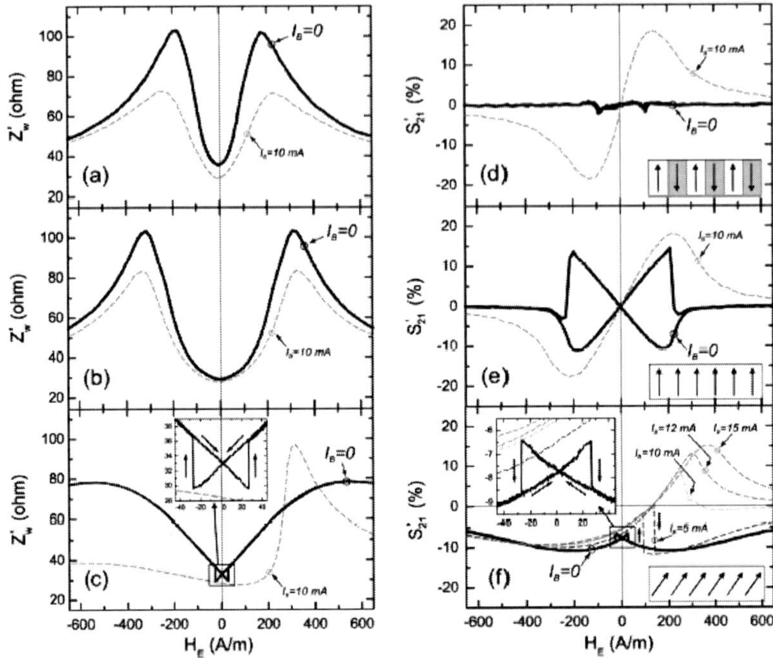

Figure 7.14 Experimental dependencies of the real parts of longitudinal impedance $Z_{zz} = Z_w/l$ (a–c) and off-diagonal impedance $Z_{\varphi z}$ (d–f) on axial magnetic field H_E at frequency $f = 30$ MHz for samples to which no stress were applied, (a) and (d), and with applied tensile, (b) and (e), and torsional, (c) and (f), stresses. The inserts show the sketch of domain structure obtained by magneto-optical polarizing microscope at $H_E = 0$ and $H_B = 0$.

When no additional stresses and no bias field are applied to the microwire, as we know from the magneto-optical experiment, the SDS is the bamboo-like (Fig. 7.12a). In this wire, the longitudinal impedance $Z_{zz}(H_E)$ (solid curve in Fig. 7.14a) exhibits a rather high symmetrical double-peak dependence with maxima at 180 A/m. The impedance changes from 35.6 to 101.5 Ωs that gives MI effect of 185%. On the other hand, the observed off-diagonal impedance $Z_{\varphi z}$ (solid curve in Fig. 7.14d) is very small and irregular, that can be expected for a wire with bamboo-like SDS in which the contribution of the domains with opposite magnetization are subtracted in the voltage induced in the pick-up coil. When the DC bias field H_B (created by DC current $I_B = 10$ mA) is applied (dashed curves in Figs. 7.14a,d), the longitudinal impedance Z_{zz} dependence demonstrates a lower sensitivity to magnetic field H_E with maxima increasing to 230 A/m. The effect of bias field H_B is much more pronounced for the off-diagonal impedance, it becomes much higher with high degree of anti-symmetry that is a result of a single domain state formation in the surface layer of the sample.

Sample 2 was soldered in a slightly pulled state that resulted in the appearance of a tensile stress in the sample and transformation of the bamboo-like SDS to the mono-domain one (Fig. 7.12b). The longitudinal impedance $Z_{zz}(H_E)$ measured at zero bias field, as shown in Fig. 7.14b, exhibits a slightly higher MI effect comparing with sample 1: the impedance changes from 29.0 to 102.5 Ωs that gives MI effect of 253%. The applied tensile stress gave rise to the increase of anisotropy field H_A and, consequently, the reduced field H_E/H_A became lower and the maxima of MI curve increased up to 320 A/m. This is a manifestation of a well-known stress-impedance effect. The effect of tensile stress on the off-diagonal impedance, shown in Fig. 7.14e is even more pronounced. Now, as the SDS is a mono-domain even at zero bias field, the dependence $Z_{\varphi z}$ is considerably higher comparing with sample 1 and hysteretic.

When the DC bias field H_B is applied (dashed curves in Figs. 7.14b,e), the longitudinal MI effect, similarly to sample 1, became lower, although in a less degree, with maxima increasing to 330 A/m. The lower effect of bias field on MI is also related to the increased anisotropy field as the reduced bias field $H_B = H_A$ decreases. After the application of bias field H_B, the off-diagonal impedance, as in sample 1, becomes higher and anti-symmetric.

Sample 3 was twisted when being soldered that induced a torsional stress in the sample and transformed the SDS into mono-domain one with helical anisotropy easy axis (Fig. 7.13a). Figures 7.14c,f show the impedance components Z_{zz} and $Z_{\varphi z}$ of the sample. Earlier, we have investigated the MI effect in microwire with helical anisotropy and found the similar MI behavior: hysteresis of MI curves, which become anhysteretic and asymmetric when the bias field is applied. As in sample 2, $Z_{\varphi z}$ is high when $H_B = 0$ that is caused by a mono-domain state. The application of axial magnetic field H_E makes the magnetization rotate (Figs. 7.13b,c) and, consequently, the impedance changes. The dashed curves show the effect of DC bias field. When a sufficiently high bias field H_B is applied (I_B = 10 mA or more), the dependence becomes anhysteretic and asymmetric. Contrary to the previous two samples, the induced voltage is not zero as $\varphi_0 \neq 0$ at H_E = 0. It is required to apply H_E as high as 125 A/m to make the magnetization lies in the transversal plane (S_{21} crosses zero) that is caused by helical anisotropy. From the measurement shown in Fig. 7.14f, we found the following parameters of the third sample: H_A = 265 A/m, α = 35° and H_B = 155 A/m with I_B = 10 mA.

References

1. Zhukova V, Chizhik A, Zhukov A, Torcunov A, Larin V, and Gonzalez J (2002). *IEEE Trans. Magn.*, **38**, 3090.

2. Chizhik A, Blanco JM, Zhukov A, Gonzalez J, Garcia C, Gawronski P, and Kulakowski K (2006). *IEEE Trans. Magn.*, **42**, 3889.

3. Panina LV and Mohri K (1994). *Appl. Phys. Lett.*, **65**, 1189.

4. Knobel M, Gomez-Polo C, and Vazquez M (1996). *J. Magn. Magn. Mater.*, **160**, 243.

5. Velásquez J, Vazquez M, and Zhukov A (1996). *J. Mater. Res.*, **11**, 2499.

6. Cobeno AF, Zhukov A, Blanco JM, and Gonzalez J (2001). *J. Magn. Magn. Mater.*, **234**, L359.

7. Usov NA, Antonov AS, and Lagarkov AN (1998). *J. Magn. Magn. Mater.*, **185**, 259.

8. Gomez-Polo C, Vazquez M, and Knobel M (2001). *Appl. Phys. Lett.*, **78**, 246.

9. Chizhik A, Gonzalez J, Zhukov A, and Blanco JM (2003). *Appl. Phys. Lett.*, **82**, 610.

10. Zhukov A, Gonzalez J, Blanco JM, Prieto MJ, Pina E, and Vazquez M (2000). *J. Appl. Phys.*, **87**, 1402.

11. Bertotti G (1998). *Hysteresis in Magnetism* (Academic Press, San Diego, USA).

12. Chizhik A, Blanco JM, Zhukov A, Gonzalez J, Garcia C, Gawronski P, and Kulakowski K (2006). *IEEE Trans. Magn.*, **42**, 3889.

13. Gomez-Polo C, Duque JGS, and Knobel M (2004). *J. Phys. Condens. Matter.*, **16**, 5083.

14. Sandacci SI, Makhnovskiy DP, Panina LV, Mohri K, and Honkura Y, *IEEE Trans. Magn.*, **35**, 3505.

15. Ipatov M, Zhukova V, Zhukov A, Gonzalez J, and Zvezdin A (2010). *Phys. Rev. B*, **81**, 134421.

Chapter 8

Helical Magnetic Structure

8.1 Introduction

The interest to the GMI effect is related with the high sensitivity of the impedance to an applied magnetic field. Earlier, it has been demonstrated [1, 2] that the application of tensile or torsion stresses results in significant changes in the GMI effect. The discovery of the torsion impedance effects [3, 4] showing high sensitivity of the impedance to the applied torsion stress attracted researchers' attention to the investigation of the helical magnetic structure in amorphous wires [5–8]. Particularly, interest to the study of the surface helical structure is determined by the fact that the glass coating sheath introduces additional stress in the surface area of the wire due to the difference between the thermal expansion coefficients of the glass coating and the metallic nucleus.

8.2 Magneto-Optical Determination of Helical Magnetic Structure

Glass-covered amorphous microwires exhibiting nearly zero negative magnetostriction, of nominal composition $Co_{69.5}Fe_{3.9}Ni_1$

Magnetic Microwires: A Magneto-Optical Study
Alexander Chizhik and Julian Gonzalez
Copyright © 2014 Pan Stanford Publishing Pte. Ltd.
ISBN 978-981-4411-25-7 (Hardcover), 978-981-4411-26-4 (eBook)
www.panstanford.com

$B_{11.8}Si_{10.8}Mo_2$ (Sample 1, metallic nucleus radius 9.5 µm, glass coating thickness 2.6 µm) and $Co_{67}Fe_{3.85}Ni_{1.45}B_{11.5}Si_{14.5}Mo_{1.7}$ (Sample 2, metallic nucleus radius 11.2 µm, glass coating thickness 3 µm) have been studied. The tensile stress up to 150 MPa and torsion stress up to $\pm40\pi$ rad m^{-1} have been applied during the experiments.

The transformation of the surface hysteresis is observed under the torsion stress (Fig. 8.1, Sample 1). The increase of the torsion stress causes the decrease of the value of the coercive field and the increase of the jump of Kerr effect signal, i.e., the increase of the jump of the circular magnetization ΔM_{CIRC}. The application of the negative torsion stress induces the change of the direction of rotation of the magnetization (Figs. 8.1a,e).

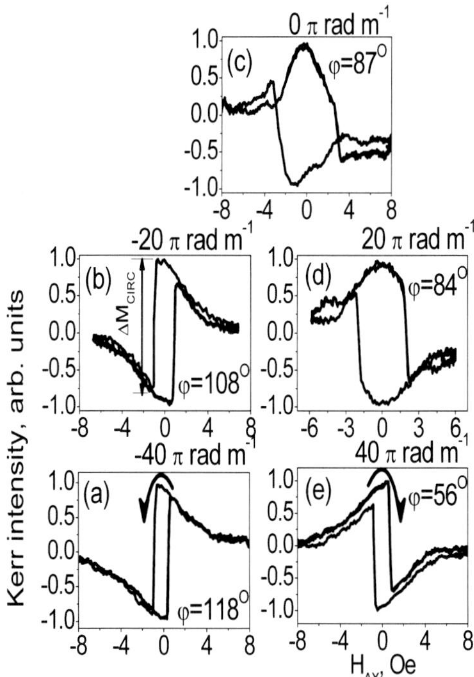

Figure 8.1 Kerr effect dependencies in the presence of torsion stress.

The application of tensile stress also induces a transformation of the surface hysteresis loop (Fig. 8.2, Sample 2). In the absence of tensile stress, the Kerr signal is almost lacking (Fig. 8.2a). The tensile stress induces the growth of Kerr amplitude and, consequently, the increase of the transverse magnetization in the surface area of the wire. Finally, Fig. 8.2c presents the typical behavior of a helical domain structure under an axial field: the monotonic increase of the Kerr signal related to the rotation of magnetization from the axial to the circular direction and a rather sharp jump of the signal, followed by the change of the sign of the signal, that is associated with the jump of the circular magnetization.

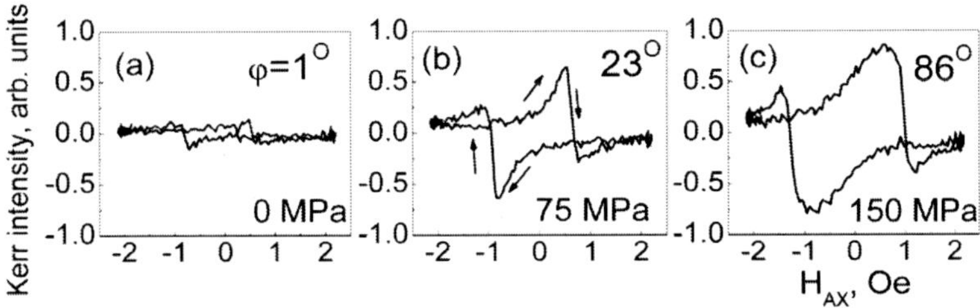

Figure 8.2   Kerr effect dependencies in the presence of tensile stress.

We consider that the observed transformation of the surface hysteresis curves is related to the formation and existence of helical magnetic structure in the studied microwires [4]. The calculation of the hysteresis loops has been performed taking into account the existence of a helical magnetic anisotropy in the microwire. The expression of the energy of system has the form

$$U = -K_U\cos^2(\theta - \varphi) - h \cdot M_S = -K_U\cos^2(\theta - \varphi) - H_{AX}\cos(\theta), \qquad (8.1)$$

where H_{AX} is the applied magnetic field, K_U is the uniaxial anisotropy constant, M_S is the saturation magnetization, φ is the angle between the anisotropy axis and the microwire axis and θ is the angle between the magnetic moment and the wire axis.

Series of the surface hysteresis curves for different values of the helical anisotropy angle φ have been obtained as the result of the numerical analysis of the equation. Figure 8.3 presents the dependence of the calculated jump of the circular magnetization ΔM_{CIRC} (see Fig. 8.1b) on the angle φ.

The comparison of the experimental and calculated hysteresis loops was performed taking into account the value and position of the jump of the circular magnetization ΔM_{CIRC}. The results of the comparative analysis are presented in Fig. 8.4 as a dependence of the angle of the helical anisotropy on the value of the torsion

(Fig. 8.4a) and tensile (Fig. 8.4b) stress. The values of the angle φ are also presented in Figs. 8.1 and 8.2. The torsion stress tilts the direction of the anisotropy toward the axial direction, while the tensile stress tilts it toward the circular one. The change of the sign of the tensile stress changes the direction of the helical curling.

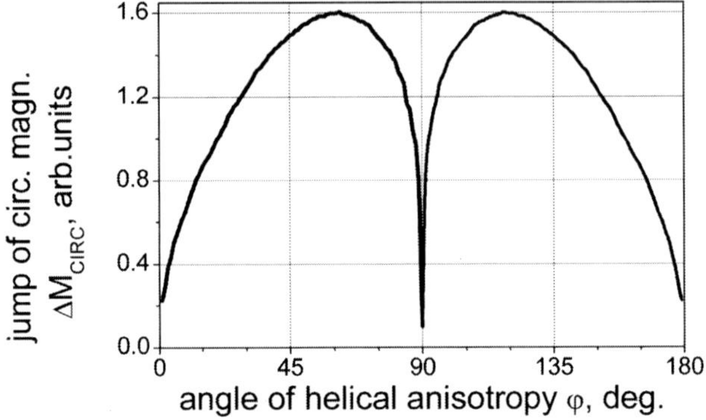

Figure 8.3   Calculated dependence of the jump of the circular magnetization ΔM on the angle of helical anisotropy φ.

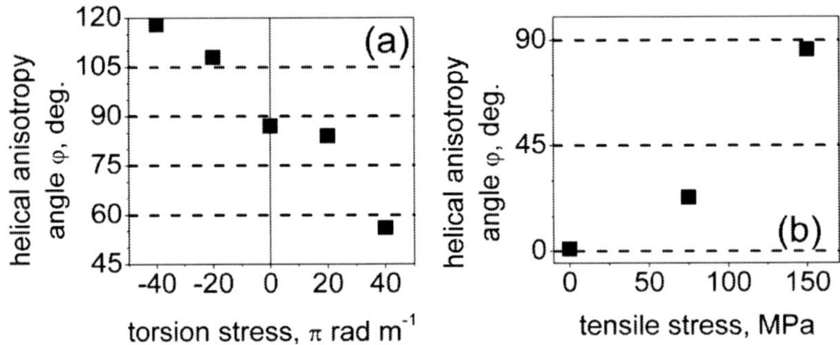

Figure 8.4   Helical anisotropy angle dependencies on torsion (a) and tensile (b) stresses obtained from the comparison of experimental and calculated hysteresis loops.

The analysis of the torsion stress experiments is in agreement with the rotational model [1], which takes into account the appearance of helical anisotropy induced by applying torsional stress. According to this model, the torsion stress induces an easy axis at the angle of $\pi/4$ with respect to the circumferential direction. Our experiments could be considered as an independent confirmation of this model, because the experimentally obtained inclination of the helical anisotropy induced by the strong enough torsion stress does not exceed the value of 45° (Fig. 8.4a).

The influence of the torsion or tension stress on the surface magnetization reversal in the Co-rich amorphous microwires has been studied. The values of the angle of the helical anisotropy in the surface area of the microwires have been determined. The torsion stress causes the decrease of the value of the surface coercive field and the increase of the jump of the surface circular magnetization. The direct observation of the tensile stress–induced growth of the transverse magnetization has been performed in the surface area of the wire. This growth could be related to the appearance of the tensile stress–induced circular anisotropy. The torsion stress tilts the direction of the anisotropy toward the axial direction, while the tensile stress tilts it toward the circular one. The change of the sign of the tensile stress changes the direction of the helical curling. The inclination of the helical anisotropy induced by the torsion stress does not exceed the value of 45°.

8.3   Transverse Kerr Effect Dependencies

Figure 8.5 presents the transverse Kerr effect dependencies on the AC electric current flowing along the wires in the presence of an DC axial bias magnetic field and with the torsion stress as a parameter. When the DC axial magnetic field is absent (Figs. 8.5a,d,i), the shape of the circular hysteresis loop is rectangular. It is related to the circular magnetic bistability in the form of large Barkhausen jump between two states with opposite direction of the surface circular magnetization. There are three different types of behavior depending on the value of the torsion stress. When the torsion stress is absent (Figs. 8.5a–c), the application of the DC axial magnetic field causes the symmetrical change of the switching field, H_{SW}, (associated with the switching current). The value of the jump of the Kerr intensity (circular magnetization) also decreases with DC axial magnetic field.

The shift of the loop is not observed. For the torsion stress of 20π rad m^{-1} (Figs. 8.5d–f), a bias field–induced shift of the loop along the "X" axis takes place. The jump of the Kerr intensity decreases with bias field and disappears when the bias field is high enough. For the torsion stress of 40π rad m^{-1} (Figs. 8.5i–k), we can see the field-induced decrease of the switching field and the shift of the loop. The main feature of this experiment is the following. The shape of the loop remains rectangular in the presence of the bias field up to the abrupt disappearance of the hysteresis loop. The shape of the loops presented in Figs. 8.5j,k is related to existence of the meta-stable state with the magnetization tilted from the circular direction, which was discussed in our paper [9].

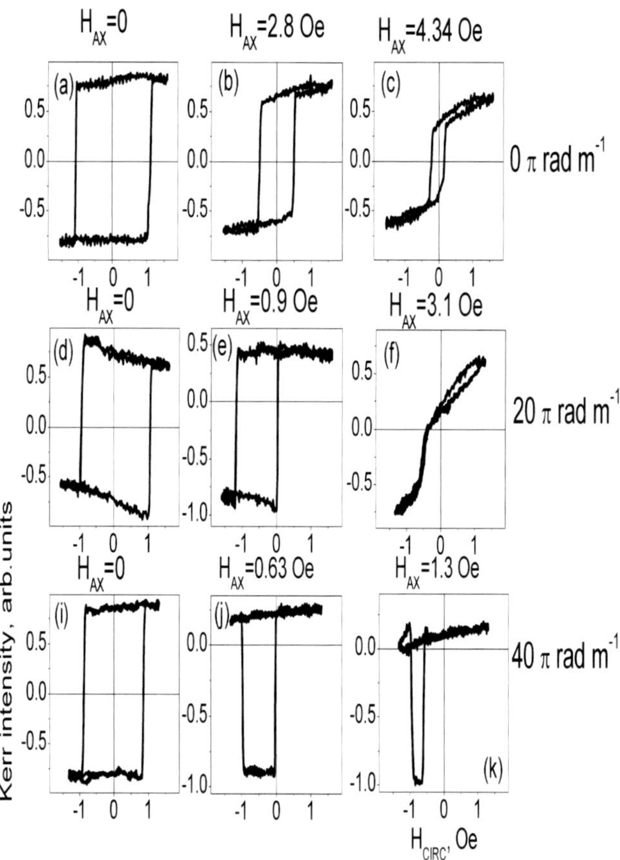

Figure 8.5   Transverse Kerr effect dependencies on circular magnetic field for different values of torsion stress with the axial bias field as a parameter. (a–c) Torsion stress 0π rad m^{-1}, (d–f) torsion stress 20π rad m^{-1}, (i–k) torsion stress 40π rad m^{-1}.

The result of the numerical analysis of the Eq. (8.1) is presented in Fig. 8.6. There is the calculated dependence of the jump of the circular magnetization ΔM_{CIRC} on the angle of helical anisotropy. Also, the series of the surface hysteresis curves for different values of the helical anisotropy angle (Fig. 8.1) have been obtained as a result of the calculation. For the angles close to 90°, the magnetization reversal is determined mainly by a fluent rotation of the magnetization and this rotation continues after the moment when the magnetization reaches the exact circular direction. For the angles close to 60°, the jump of the magnetization occurs at the moment when the direction of the magnetization is close to the circular magnetization. Because of that the value of the ΔM_{CIRC} is large for these angles.

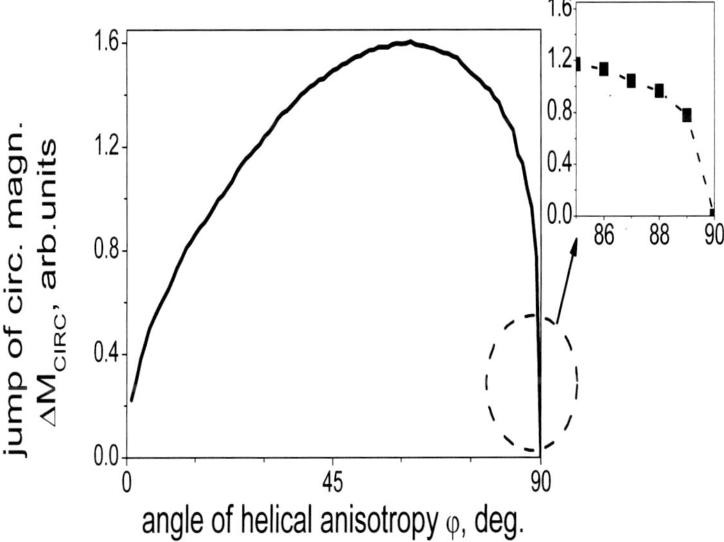

Figure 8.6 Calculated dependence of the jump of the circular magnetization ΔMCIRC on the angle of helical anisotropy φ.

The Kerr effect experimental results and the results of the calculation have been compared taking into account the value of the jump of the circular magnetization ΔM_{CIRC}. This jump is related to the overcoming of the helical hard axis.

The experimental results presented in Fig. 8.5 are in agreement with the above-performed analysis of the magnetic structure in the surface area of the studied microwire. The observed shift of the hysteresis loops means that the sensitivity of the studied system to the DC axial bias field grows with the increase of the inclination of the helical anisotropy from 90°.

The shape of the hysteresis loop presented in Fig. 8.5c means that the DC axial field inclines the magnetization toward the axial direction in the surface area of the microwire. Therefore, the jump happens between the conditions with the inclined magnetization. The shift of the hysteresis loop reflects the action of the bias field on the nucleation process of the circular domains. For the helical anisotropy inclined enough from the circular direction, the DC axial field favors and delays the nucleation of the domains with different directions of the magnetization. This effect becomes stronger as the inclination increases.

For the torsion stress of 40π rad m^{-1}, the strong rectangular shape of the hysteresis is observed up to the abrupt disappearance along with the shift of the loop. In the presence of the DC axial field the magnetization reversal occurs in the form of large Barkhausen jump between two states with helical direction of the magnetization. For the torsion stress of 40π rad m^{-1}, the appearance of the hysteresis loop has a threshold character. The observed behavior could be considered as the confirmation of the torsion stress–induced growth of the value of the helical anisotropy in the surface area of the studied microwire. For the relatively weak helical anisotropy (Figs. 8.5c,f), the fluent rotation of the magnetization takes place along with the jump of the magnetization, while for the strong helical anisotropy the rotation is lacking and only the sharp jump of the surface magnetization is observed (Fig. 8.5k).

8.4 Helical Magnetic Structure in Microwires with Different Value of Geometric Ratio

Figure 8.7 presents the transverse Kerr effect dependencies on the AC circular magnetic field (frequency $f = 60$ Hz) with the DC axial magnetic field as a parameter (nominal composition $Co_{69.5}Fe_{3.9}Ni_1B_{11.8}Si_{10.8}Mo_2$, $\rho = 0.785$ and 0.930.

Figures 8.7a–e,f–k show the axial field–induced transformation of the surface circular hysteresis loop for the microwires with $\rho = 0.93$ and 0.785, respectively. Figure 8.8 presents the transverse Kerr effect dependencies on the AC axial magnetic field (frequency $f = 60$ Hz) for the microwires with $\rho = 0.93$ and 0.785.

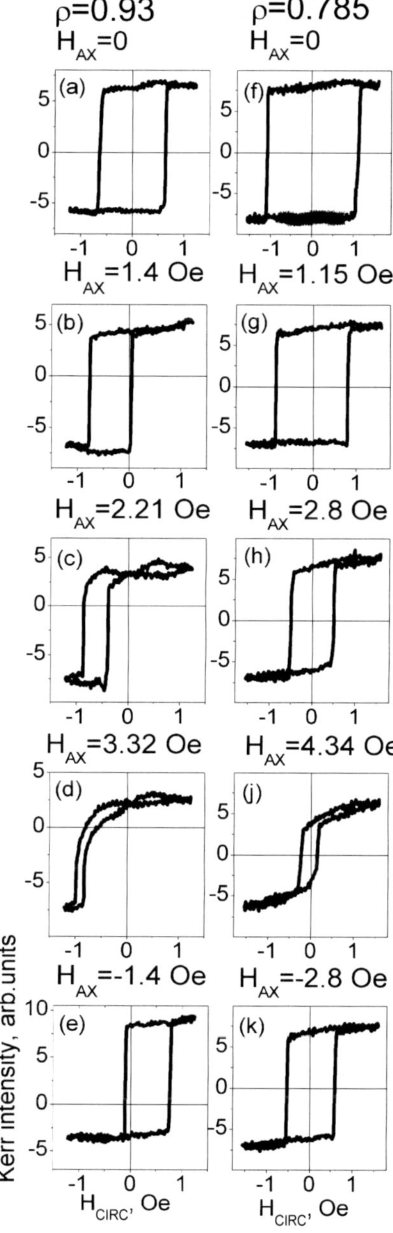

Figure 8.7 Transverse Kerr effect dependencies on circular magnetic field with axial bias field as a parameter. (a–e) Wire with ρ = 0.93, (f–k) wire with ρ = 0.785.

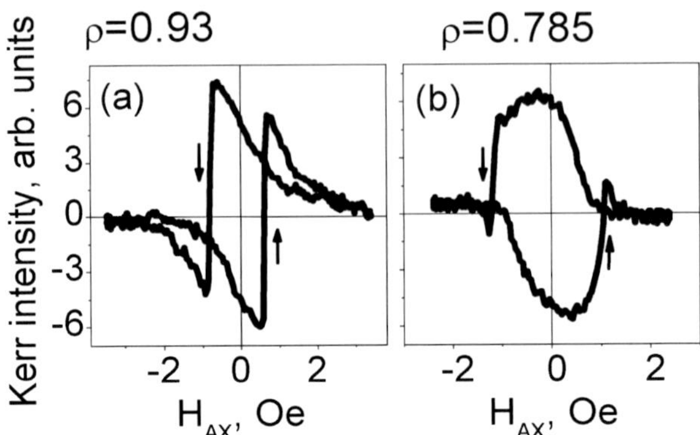

Figure 8.8 Transverse Kerr effect dependencies on axial magnetic field for two wires with different geometric ratio ρ.

When the DC axial magnetic field is absent, the shape of the circular hysteresis loop is perfectly rectangular (Figs. 8.7a,f), that is related to the circular magnetic bistability. For the circular magnetic field smaller than the value of the circular coercive field, the hysteresis loop is not observed (see the short black line in Fig. 8.7f).

The DC axial magnetic field initiates the transformation of the circular hysteresis loop, and this transformation occurs in two different ways for the two studied microwires. Let us consider first the transformation of the hysteresis loop for the microwire with ρ = 0.93 (Figs. 8.8a–e). The application of the DC axial magnetic field causes the asymmetrical change of the switching field, H_{SW}, (associated with the switching current) (Figs. 8.7b–d). We can see that the value of one of the switching field (H_{SW}^{+}) decreases when the value of another switching field (H_{SW}^{-}) increases. This effect manifests as the observed "shift" of the hysteresis loop along the X axis.

The transformation of the circular hysteresis loop for the microwire with ρ = 0.785 occurs in another way. The absolute values of H_{SW}^{-} and H_{SW}^{+} decrease symmetrically and the "shift" effect is not observed (Figs. 8.7g–j). The application of the negative axial magnetic field transforms the hysteresis loop in the same way as the positive one.

A great difference is seen also between the hysteresis loops obtained in the AC axial magnetic field for two studied microwires (Figs. 8.8a,b). When the external axial magnetic field increases, a monotonic increase of Kerr signal is observed. This increase is related to the rotation of the magnetization from the axial to the circular direction in the outer shell of the wire. After that, a sharp jump of the signal takes place with the change of the signal sign. This is related to the jump of the circular magnetization. For the microwire with ρ = 0.93, this jump occurs at relatively small axial field and the value of the jump of magnetization in high enough. For the microwire with ρ = 0.785, the magnetization reversal consists mainly of the fluent rotation of the magnetization. The jump of the circular magnetization is small.

Following the Eq. (8.1) we obtain calculated hysteresis curves of two types: the dependencies of the circular magnetization on the circular magnetic field in the presence of the axial bias field and the dependencies of the circular magnetization on the axial magnetic field.

The results of the numerical analysis of the Eq. (8.1) are presented in Figs. 8.9 and 8.10. There are the hysteresis loops for two configurations: when the direction of the uniaxial anisotropy is close to the circular direction in the microwire (Figs. 8.9d–f, 8.10b) and when this direction is noticeably inclined from the circular direction (Figs. 8.9a–c, 8.10a).

The results on the calculated hysteresis loops allow us to conclude as follows. The presence of the anisotropy inclined from the circular direction causes an asymmetric transformation of the hysteresis loop. The shift of the curve takes place in the presence of the axial bias field (Figs. 8.9a–c). The direction of the shift depends on the sign of the bias field. When the anisotropy direction is directed exactly along the circular direction, the bias field–induced shift is not observed and the value of the switching field decreases symmetrically. For the case when the angle of anisotropy φ is close to 90°, the shift is very small (Figs. 8.9 d–f).

The calculated loops presented in Fig. 8.10 demonstrate how the direction of the anisotropy influences on the dependence of the circular magnetization on the axial magnetic field. When the direction of the anisotropy is close to the circular direction (Fig. 8.10b), the magnetization reversal consists mainly of the rotation of the magnetization. The sharp jump of the magnetization is related to the overcoming of the hard axis close to the axial direction. (When the anisotropy is directed exactly along the circular direction, the jump is not observed and only the rotation of the magnetization takes place.) The inclination of the direction of the magnetization toward the axial direction causes an increase of the jump of the circular magnetization and a decrease of the part of hysteresis related to the rotation of the magnetization. Therefore, the observed features of the calculated hysteresis curves are very similar to the features observed in the experimental Kerr effect curves and, consequently, we can conclude that these experimentally observed properties are related to the different direction of the anisotropy axis in the surface area of the wire or, in another words, to the degree of the helical anisotropy: For the wire with $\rho = 0.785$, the direction of the surface anisotropy is very close to the circular direction, and for the wire with $\rho = 0.93$, this direction is inclined toward the axial direction.

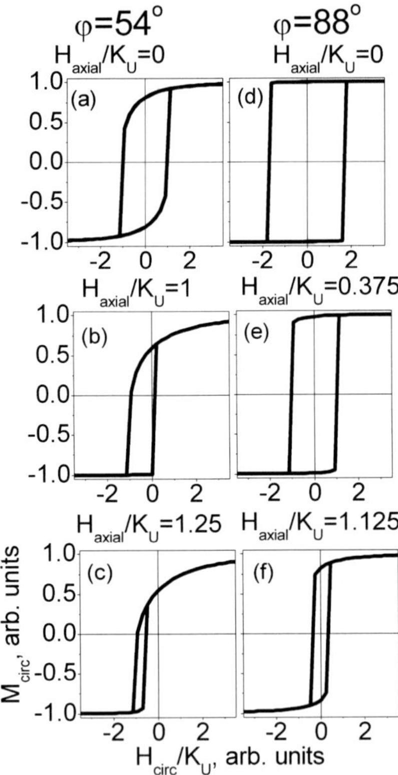

Figure 8.9 Calculated dependencies of circular magnetization on normalized circular magnetic field with normalized axial bias field as a parameter for two different angles of anisotropy direction φ. (a–c) $\varphi = 54°$, (d–f) $\varphi = 88°$.

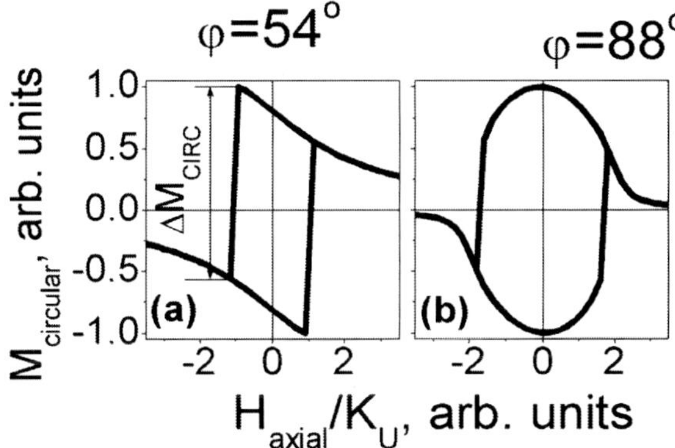

Figure 8.10   Calculated dependencies of circular magnetization on normalized axial magnetic field for two different angles of anisotropy direction φ. (a) $\varphi = 54°$, (b) $\varphi = 88°$.

8.5 Surface and Bulk Magnetic Hysteresis Loops of Co-Rich Glass–Covered Microwires

The bulk hysteresis loops obtained for the two studied wires in axial magnetic field are presented in Figs. 8.11a,b. Just as the experimental MOKE curves, these loops also have the parts related to the rotation of the magnetization and jumps of the magnetization.

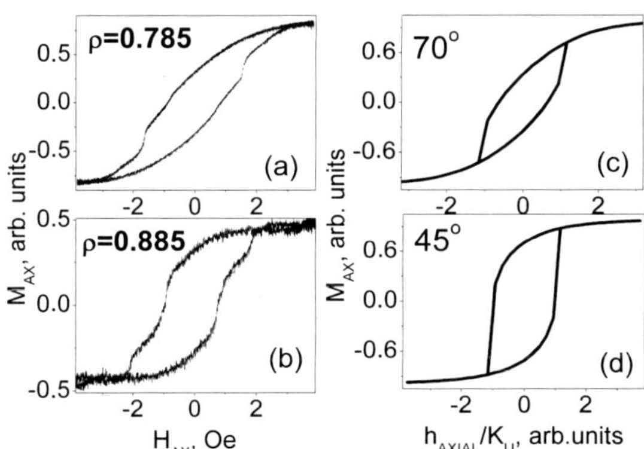

Figure 8.11   (a, b) Experimental bulk dependencies on axial magnetic field for the two wires with different thickness of glass covering. (c, d) Calculated dependencies of axial magnetization on normalized axial magnetic field for two angles of anisotropy direction φ.

The transverse Kerr effect curves and the conventional curves reflect the change of the circular and axial projection of the magnetization, respectively. Therefore, they could be considered as complement date. From another side, the Kerr effect loops contain information about the surface magnetization reversal when the conventional loops present the magnetization reversal of the whole volume of the wire. Consequently, this complement could not be considered as a complete one.

Frequency dependence of the coercive field has been studied for the wire with ρ = 0.885 using the conventional and Kerr effect techniques. The results of the conventional experiments are presented in Fig. 8.12. Frequency dependence of the coercive field H_C has been analyzed using the calculations presented in Ref. [10]. It was shown that the result of the solution of the equation of domain wall motion in the following relation has been obtained [10]:

$$H_{CD} = H_{CO} + 4fH_0(L + 2I_SA)/K, \tag{8.2}$$

where H_{CD} is the dynamic switching field, H_{CO} is the static coercive field, f is the frequency of the magnetic field, H_0 is the amplitude of the magnetic field, L is the damping coefficient, I_S is the saturation magnetization, K is the elastic coefficient, and A is a proportionality constant. This linear dependence of the coercive field on frequency could be described in terms of domain nucleation mechanism. The nucleation process is associated with the overcoming of the energy barrier.

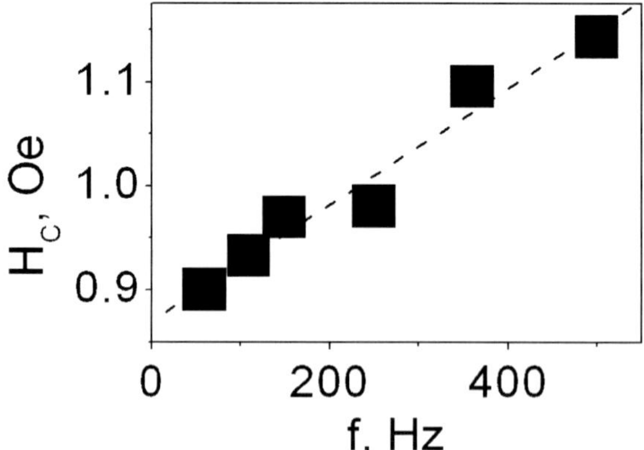

Figure 8.12 Dependence of the volume coercive field on the frequency of magnetic field for the wire with ρ = 0.885.

Thermoactivation mechanism of the overcoming was satisfactorily employed for the explanation of the coercive field fluctuations observed in amorphous wires. Frequency of the applied field in this way affects the value of the coercive field. The experimental dependence of the coercive field on the frequency fits well the linear function that gives us possibility to conclude that the above mechanism of magnetization reversal could be applied to this experiment.

Kerr effect dependencies also changed when the frequency of the axial magnetic field increased. But the transformation of the surface curves differs from the transformation of the bulk curves. Figure 8.13 demonstrates the observed decrease of $H_{C\text{-MO}}$ and the increase of I_{MAX} ($H_{C\text{-MO}}$ should be considered as the field at which the drastic change of the circular magnetization starts, I_{MAX} is the maximal intensity of the Kerr signal during the magnetization reversal). Taking into account that the Kerr intensity is proportional to the transverse magnetization in the surface area of the wire, $I_{MAX} \sim M_{MAX}$, where M_{MAX} is the maximal value of the transverse (circular) magnetization.

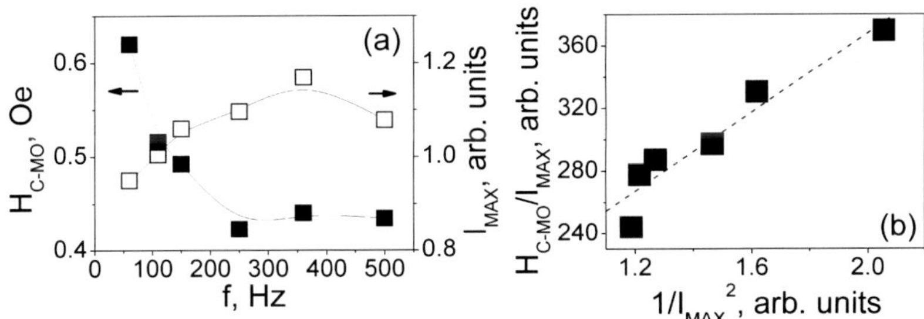

Figure 8.13 (a) $H_{C\text{-MO}}$ and I_{MAX} dependencies on frequency of magnetic field for $\rho = 0.885$, (b) $H_{C\text{-MO}}/M_{MAX}$ dependence on $1/M_{MAX}^2$.

Analyzing the results obtained in the Kerr effect experiments we consider that the magnetization reversal process in the surface area of the wire occurs as the formation of a circular domain in the outer shell. The coercive field $H_{C\text{-MO}}$ is determined by the circular domain nucleation. The expression for $H_{C\text{-MO}}$ can be presented as [11]

$$H_{C\text{-MO}} = \alpha/(M_{MAX}-N\,M_{MAX}), \tag{8.3}$$

where the first term is related to losses of energy at nucleation process, and the second term to the demagnetizing field of the nucleus (N is the demagnetizing factor). We consider, within numerical factor, that $\alpha = \sigma/v^{1/3}$, where σ is energy of domain wall, v is critical volume of the nucleus.

To verify the above-mentioned assumption regarding the mechanism of the magnetization reversal in the outer shell of the microwire, the $H_{\text{C-MO}}/I_{\text{MAX}}$ dependence on $1/I_{\text{MAX}}^2$ has been plotted (Fig. 8.13b). Taking into account that $I_{\text{MAX}} \sim M_{\text{MAX}}$, this dependence could be considered as the analogy of $H_{\text{C-MO}}/M_{\text{MAX}}$ $(1/M_{\text{MAX}}^2)$ dependence. Good fitting of the experimental points by the linear function demonstrates the strong relation between the $H_{\text{C-MO}}$ and the M_{CIRC}.

The strong difference of the frequency dependence of the H_{C} and $H_{\text{C-MO}}$ means that the magnetization reversal happens independently enough in the surface and in the volume of the wire. $H_{\text{C-MO}}$ is smaller than H_{C}. Therefore, the nucleation starts in the surface in the moment when the circular projection of the magnetization reaches the sufficient value. This value of this circular projection is the key parameter that changes with the magnetic field frequency. We suppose that the one of the possible reasons of this change is the following one. The increase of the frequency means the growth of the velocity of the increase of the magnetic field dH/dt. Therefore, the magnetic field increases more quickly for the higher frequency. The quicker increase of magnetic field causes the quicker increase of the circular magnetization. Therefore, for the higher frequency, the magnetization could rich the larger value in the moment when the domain nucleus becomes the stable one, i.e., in the moment close to the nucleation. Finally, the relation of the dH/dt and the probability of the nucleation realizes in the decrease of the $H_{\text{C-MO}}$ with frequency.

We consider that the surface magnetization reversal is determined mainly by the nucleation mechanism. The sharpness of the ΔM_{CIRC} jump and Kerr microscopy study of the magnetization reversal in amorphous wires [12] permit us to make such conclusion.

The Eq. (8.3) is used in general for the temperature dependence of the coercive field. We have used it for the frequency dependence of the coercive field and we believe that this application is reasonable taking into account the similar character of the influence of the temperature and frequency on the probability of the domain nucleation.

For the determination of the angle of helical anisotropy, we applied the calculation that did not include the domain wall motion, while for the frequency experiments, we use the equation that takes into account the nucleation and domain wall motion. For the studied case, the shape of the hysteresis curve and particularly the value and the position of the ΔM_{CIRC} jump, the value that we used in the analysis, are determined mainly by the direction of the helical anisotropy and to a lesser degree, by the details of the domain wall motion. However, for the dependencies of the coercive field on the frequency, the features of the magnetization reversal become important, because of the change of the frequency of the external magnetic field influences strongly on the mechanism of the magnetic reversal in the wires.

8.6 Experimental Determination of Limit Angle of Helical Anisotropy in Amorphous Magnetic Microwires

The transformation of the surface hysteresis loops has been produced by the external torsion stress. The most important feature of this transformation is the stress-induced change of the value and direction (sign) of the jump of the Kerr intensity ΔI (the circular magnetization ΔM_{CIRC}). For the torsion stress of -2.2π rad m^{-1} value (Fig. 8.14b), there is no jump of the circular magnetization. The increase of the applied stress of negative value (-22π rad m^{-1}) causes the appearance of the ΔM_{CIRC} of the opposite sign (Fig. 8.14a).

The observed transformation is related to the torsion-induced change of the surface helical anisotropy. The pictures presented in Fig. 8.15 demonstrate schematically the change of the direction of the surface helical anisotropy.

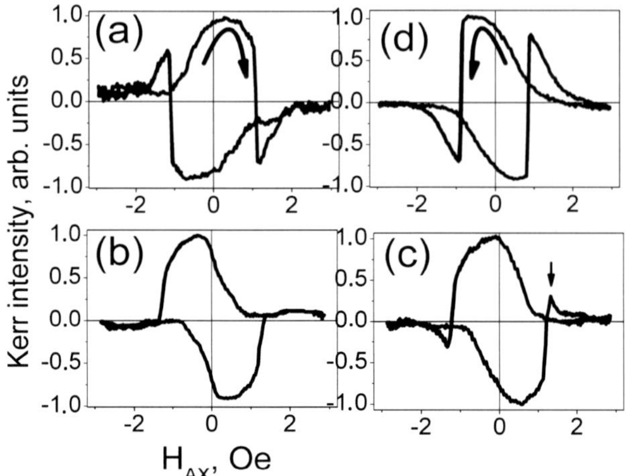

Figure 8.14   TMOKE dependencies in the presence of torsion stress. (a) $\tau = -22\pi$ rad m^{-1}, (b) $\tau = -2.2\pi$ rad m^{-1}, (c) $\tau = 0$, (d) $\tau = 8.9\pi$ rad m^{-1}. Nominal composition Co$_{69.5}$Fe$_{3.9}$Ni$_1$B$_{11.8}$Si$_{10.8}$Mo$_2$, metallic nucleus diameter 19 μm, glass coating thickness 2.6 μm.

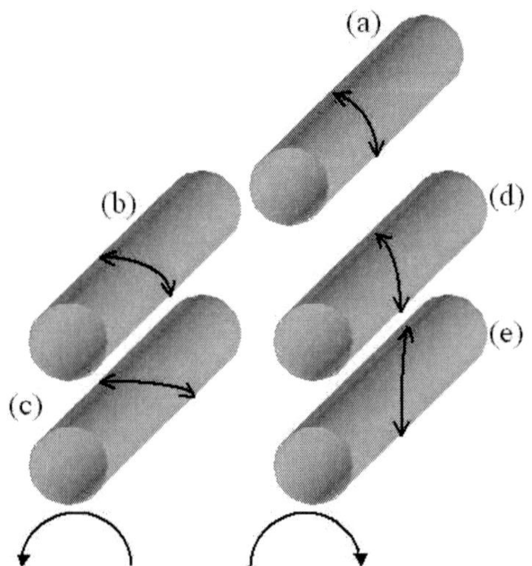

Figure 8.15   Schematically pictures of the inclination of the axis of helical anisotropy induced by the torsion stress. (a) without stress; (b), (c) left torsion; (d), (e) right torsion.

The Kerr intensity is proportional to the transverse magnetization in the surface area of the wire, $\Delta I/I_{MAX} \sim \Delta M_{CIRC}/$

M_{MAX}, where M_{MAX} is the maximal value of the transverse (circular) magnetization during the magnetization reversal. Figure 8.16 presents the experimental dependence of the normalized value of the jump of the Kerr intensity $\Delta I/I_{MAX}$ (I_{MAX} is the maximal intensity of the Kerr signal during the magnetization reversal) on the applied torsion stress. We pay attention to this parameter because the value of the jump of the circular magnetization ΔM_{CIRC} is the key parameter that permits us to establish the correlation between the shape of the surface hysteresis loop and the angle of the helical anisotropy and also to determine this angle.

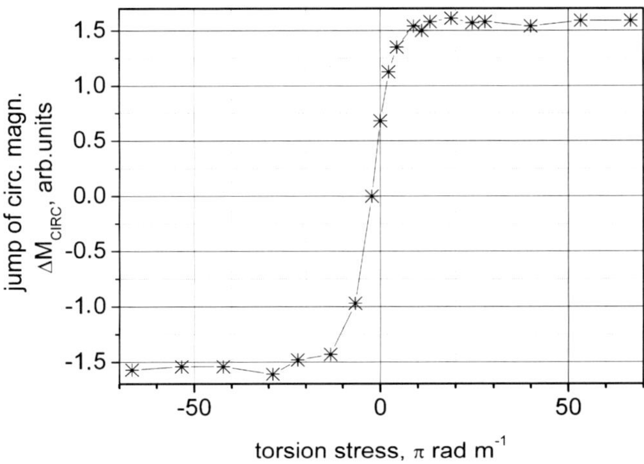

Figure 8.16 Experimental dependence of the normalized value of the jump of the Kerr intensity $\Delta I/I_{MAX}$ on applied torsion stress.

The applied magnetic field can be expressed as a superposition of two mutually perpendicular fields (H_{AX} and H_{CIRC}). The direction of the anisotropy was changed from axial to circular direction. The numerical calculation was done by the coherent rotation approach. The hysteresis loops was computed by the minimalization of the energy term of the system. The jump of the circular magnetization ΔM_{CIRC} considerably depends on the value of the angle of the helical anisotropy. Figure 8.17 shows the calculated dependence of the $\Delta M_{CIRC}/M_{MAX}$ on the angle of the helical anisotropy. The jump is equal zero for the angle of 90° (the case of the transverse anisotropy). The calculated dependence of the ΔM_{CIRC} has two maximum for the values of the angle of 62° and 118°.

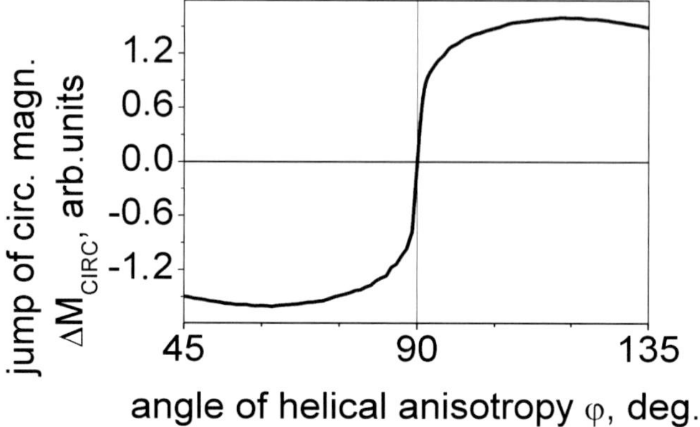

Figure 8.17 Calculated dependence of the jump of the circular magnetization $\Delta M_{CIRC}/M_{MAX}$ on the angle of helical anisotropy φ.

Figure 8.18 presents the result of the comparative analysis of the experimental results presented in Fig. 8.16 and the results of the calculations presented in Fig. 8.17. We have determined and constructed the dependence of the angle of the helical anisotropy on the applied torsion stress. In the absence of the applied stress, the anisotropy was directed almost to the transverse direction but not exactly to this one. The small peak marked in Fig. 8.14c means that the small jump of circular magnetization takes place and that the angle of anisotropy is not exactly 90°. The application of relatively small stress of -2.2π rad m^{-1} value induces the disappearance of this small peak and the transverse anisotropy but the helical one does not exist at this value of the torsion stress. The increase of the absolute value of the stress causes the growth of the absolute value of the angle of the helical anisotropy. This growth ends at the applied stress value of around $\pm$ 40π rad m^{-1}. Therefore, as it was predicted in [13, 14], the torsion stress–induced inclination of the helical anisotropy does not exceed 45° from the transverse direction in spite of the value of the applied stress was high enough.

The experimental limit values of the angle of the helical anisotropy have been obtained as the result of the analysis of the torsion stress–induced transformation of the surface magnetization reversal in the Co-rich amorphous microwires. The relation between the value of the torsion stress and the angle of helical anisotropy angle has been established. Now we have the method that permits us to present the results of the experiments with torsion stress as a dependence on the angle of helical anisotropy.

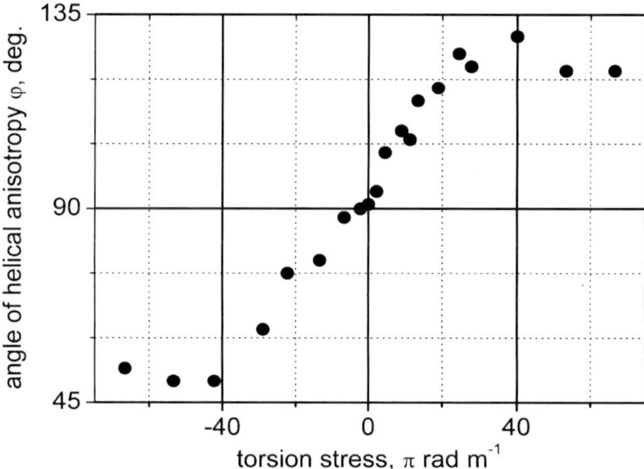

Figure 8.18 Dependence of angle of helical anisotropy on applied torsion stress.

8.7 Correlation of Magnetic Behavior with Diameter of Microwire

Figure 8.19 presents the TMOKE dependencies on the AC circular magnetic field with the DC axial magnetic field as a parameter for the wire of d = 16.8 μm. When the DC axial magnetic field is absent, the shape of the circular hysteresis loop is perfectly rectangular (Fig. 8.19) that is related to the circular magnetic bistability. For the circular magnetic field smaller than the value of the circular coercive field, the hysteresis loop is not observed. The DC axial magnetic field initiates the transformation of the circular hysteresis. The application of the DC axial magnetic field causes the asymmetrical change of the coercive field, HC (associated with the switching current I_C). We can see that the value of one of the coercive fields ($H_{C2} \propto I_{C2}$) decreases when the value of another coercive field ($H_{C1} \propto I_{C1}$) increases (or vice versa, depending on the direction of the DC field). This effect manifests as the observed "shift" of the hysteresis loop along the X axis. Figure 8.20 demonstrates the experimental dependence of switching currents I_{C2} and I_{C2} (coercive fields H_{C1} and H_{C2}) on the DC axial field.

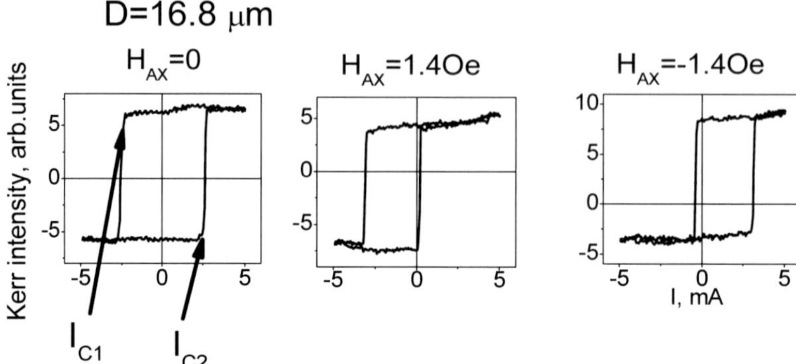

Figure 8.19 TMOKE dependencies on circular magnetic field with axial bias field as a parameter for d = 16.8 μm.

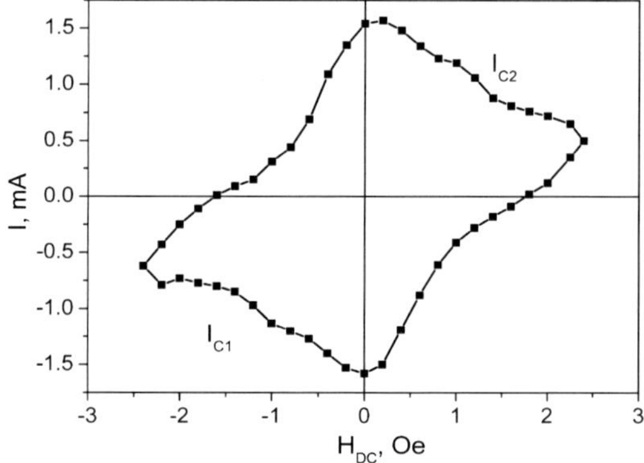

Figure 8.20 Experimental dependence of switching current on DC axial field.

We have obtained calculated hysteresis curves as dependencies of the circular magnetization on the circular magnetic field in the presence of the axial bias field. The results of the calculations are summarized in Fig. 8.21 for the angle of helical anisotropy of 72°. Good quality coincidence of the experimental results with the results of the calculation could serve as direct confirmation of the strong correlation between the direction of the helical magnetization in the outer shell and the direction of axial magnetization in the inner core in the frame of the core–shell model.

The absolute value of the H_{CIRC} decreases with DC axial magnetic because H_{AX} increases the probability of the nucleation of surface magnetic domains during magnetization reversal. As it is possible to see, the top and bottom vertices of the parallelogram property are shifted. The vertex takes place when the total field ($H_{\mathrm{CIRC}} + H_{\mathrm{AX}}$) is directed to the direction of helical anisotropy. This value of the shift is determined by the angle of helical anisotropy.

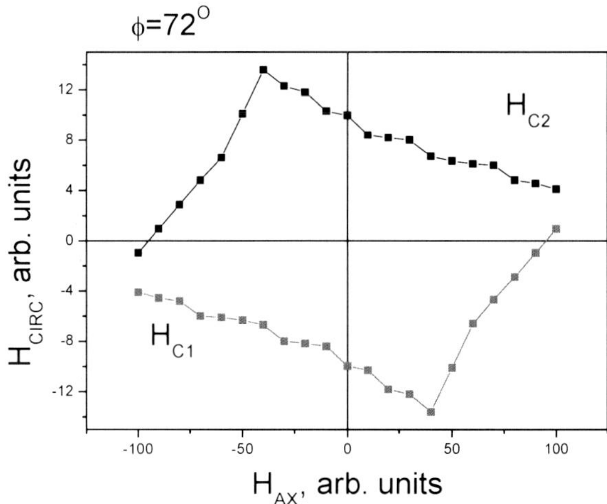

Figure 8.21   Calculated dependence of switching current on DC axial field.

Figure 8.22 shows the DC field–induced transformation of the transverse hysteresis loop for the microwire of d = 10 μm. It is possible to see the strong difference between the wires with d = 16.8 μm and d = 10 μm. The main peculiarity is that for the case of d = 10 μm, the DC field induces the decrease of H_{C1} and H_{C2} fields and small unidirectional shift of the hysteresis loop independently on the direction of the DC field. This effect has the explication in the supposition that the magnetic structure of the microwire of this diameter consists only of the helical magnetic structure with strongly determined direction of the curling of the helicality. The presented experiment could be considered as a confirmation of the prediction that the axially magnetized inner core could disappear at the determined diameter of microwire [15].

Figure 8.23 presents the hysteresis loop for the wire with d = 5.8 μm in the presence of DC axial field. The TMOKE hysteresis curve consists of two jumps related to jumps of the magnetization on the surface of the microwire. The jumps are accompanied by the local hysteresis. The positions of these jumps could be shifted by the axial magnetic field. The observed behavior is related to the existence of axial and circular anisotropies which determine two stable magnetic structures. The additionally performed fluxmetric measurements also demonstrate the "two-jump" behavior that confirms that two stable magnetic phases exist not only in surface of the microwire but also in the volume of it.

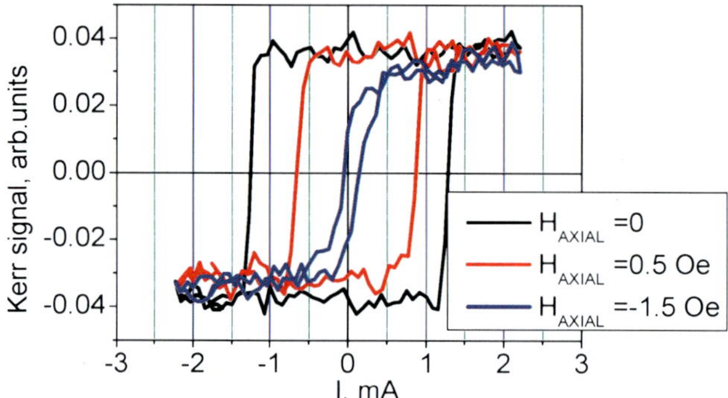

Figure 8.22 TMOKE dependencies on circular magnetic field with axial bias field as a parameter for d = 10 μm.

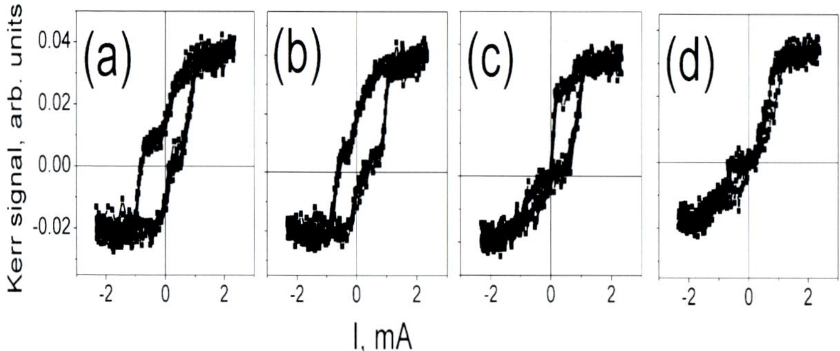

Figure 8.23 TMOKE dependencies on circular magnetic field with axial bias field as a parameter for d = 5.8 μm.

The magnetization reversal process has been studied in the series of the Co-rich amorphous glass-covered microwires with diameters of 16.8–5.8 μm. The decrease of the diameter causes the transformation of the magnetic structure in the microwire. In the microwire with the diameter of 16.8 μm, the helical magnetic structure existing in the surface has strong correlation with the axially magnetized inner core. Predicted disappearance of the axially magnetized inner core is observed for the diameter 10 μm. The TMOKE study, for the first time performed in thin microwire with diameter of about 5 μm, demonstrates that the decrease of the microwire diameter up to this value causes the formation of two stable magnetic phases both in the surface and in the volume of the microwire.

8.8 Magnetic Domain Structure Studied by Bitter Technique

Amorphous wires with nominal composition $Fe_{73.5}Si_{13.5}B_9Nb_3Cu_1$ were prepared by the rotating-water quenching method having a diameter of $2r = 133$ μm. The samples were cut into 10 cm-long pieces for the measurements. A positive saturation magnetostriction value of $\lambda_S = +18 \times 10^{-6}$ was measured by means of the small angle magnetization rotation method, and a shear modulus of 10^{-9} N m^{-2} was determined by employing a torsion pendulum.

Magnetic domains have been observed by Bitter technique in a metallographic microscope. Ferrofluid nanoparticles were supplied by Ferrotec, Inc., and observations were done under a constant applied magnetic field (24 kA m^{-1}) perpendicular to the wire axis to enhance the contrast. The experimental setup allows twisting the sample simultaneously by clamping the ends of the wire to measure the axial hysteresis loops. One of the clamps can be rotated to apply a pure torsion by manually twisting the wire around its axis; hence, such system avoids any tension along the wire.

The wire is inserted into a short glass tube (1 mm outer diameter) onto which a 2000-turn pickup coil was wounded to measure the axial magnetization by a conventional induction method at the frequency of 48 Hz. Saturation magnetization value of $\mu_0 M_S = 1.31$ T has been measured for the studied wire.

The surface domain structure of untwisted wires is shown in Fig. 8.24a. The domain patterns exhibit a maze configuration and zigzag walls on the whole surface, indicating the presence of stripe domains with unclosed magnetic flux.

Magnetization can be deduced to lay roughly perpendicular to the wire surface, as has been observed by the magneto-optical technique [16]. Furthermore, the patterns of domain walls make different angles to the wire axis denoting the presence of twisting residual stresses. The modification of domain structure with increasing applied clockwise torsion is shown in Figs. 8.24b,c. The maze domain structure is no more observed when torsion is applied to the wire, even at low values of the applied torsional strain ξ. As the applied torsion increases, the domain structure changes, showing the presence of magnetic domains and 180° walls tilting to the wire axis in a helical structure. The most interesting features of this domain structure, such as domain width and wall angle, depend on the applied twist. The effect of applying torsion in the counterclockwise sense is, as expected, that of orientating the domain walls along the opposite helix (see Fig. 8.24d). A nearly symmetric dependence of the domain width w with the applied torsion is observed (see Fig. 8.24a), with a minimum width of around 10 μm at zero applied torsion, and two maxima (w = 30 μm) at ξ = ±10 (2π rad m^{-1}). Higher values of the applied torsional strain (both clockwise and counterclockwise) lead to a decrease of the domain width. The 180° wall angle with the wire axis Ψ increases quickly with the applied torsion (see Fig. 8.25b), either

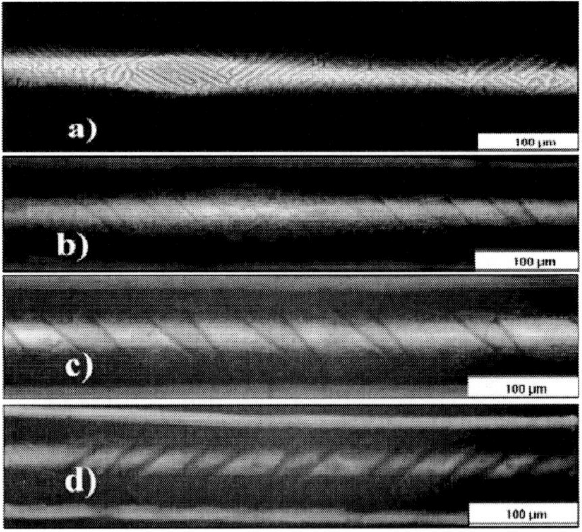

Figure 8.24 Domain structures of Fe$_{73.5}$Si$_{13.5}$B$_9$Nb$_3$Cu$_1$ amorphous wires: (a) Untwisted and under different values of the applied torsion in the clockwise sense: (b) ξ = 10 (2π rad m^{-1}), (c) ξ = 12.5 (2π rad m^{-1}), and in the counterclockwise sense: (d) ξ = −12.5 (2π rad m^{-1}).

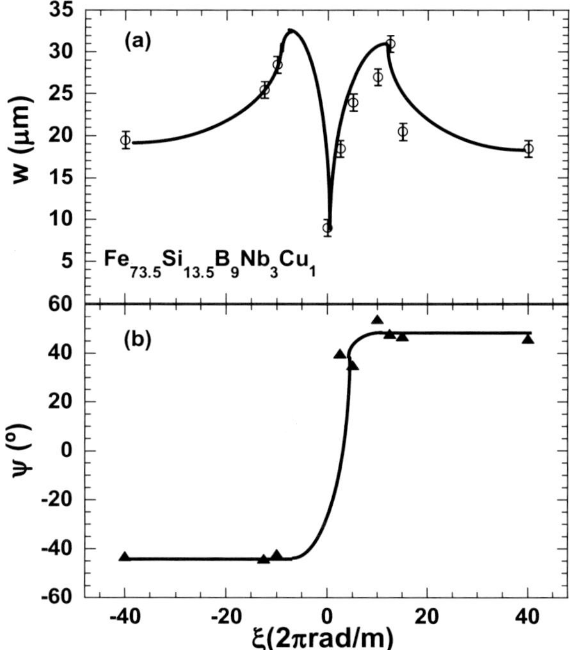

Figure 8.25   Torsional strain dependence of (a) the domain width w and (b) domain wall angle with respect to the wire axis Ψ. The lines between points are guides for the eyes.

in clockwise or counterclockwise senses, approaching to its saturation value of $\Psi = \pm 45°$ for $\xi \geq 10$ (2π rad m^{-1}). The angle Ψ for vanishing applied torsion is difficult to determine due to the presence of maze domains. The behaviors of the coercive field H_C and the ratio between remanence and saturation magnetization m_r, with applied torsion, are displayed in Fig. 8.26. The reduced remanence first increases with torsion and, after reaching a maximum for $\xi = 10$ (2π rad m^{-1}), it slightly decreases toward its saturation value at $m_r = 1/\sqrt{2}$. It must be mentioned that for m_r, there is an opposite contribution from the core and shell regions of the wire [17]. In turn, the coercive field first decreases with torsion and reaches a minimum value at around $\xi = 2.4$ (2π rad m^{-1}) and then continuously increases linearly with the square root of the applied torsion. Owing to the amorphous nature of the wire, it is expected that the magnetoelastic anisotropy plays the dominant role in the magnetization process.

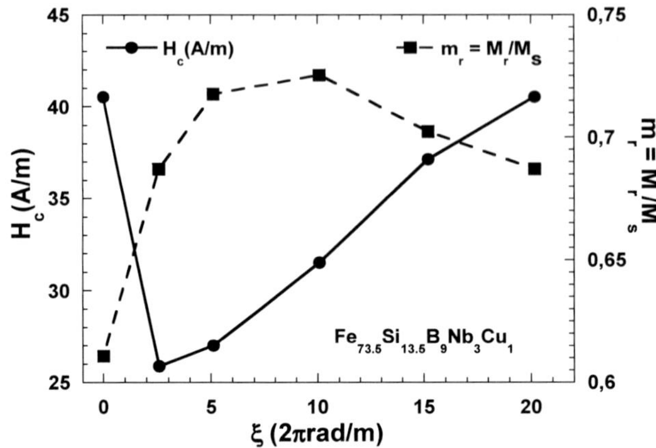

Figure 8.26 Torsional strain dependence of the coercive field H_C, (referred to down x scale) and reduced remanence m_r (referred to upper x scale). The lines between points are guides for the eyes.

In fact, the application of a torsional stress induces helical magnetoelastic anisotropy at 45° to the wire axis. This anisotropy adds to the internal magnetoelastic anisotropy induced during quenching, therefore, modifying the magnetic response of the wire. The axial magnetization process should be determined by domain wall displacements, particularly through a nucleation and depinning mechanism. In this case, and in order to observe a bistable behavior, the switching field H^* must be proportional to the energy stored in the propagating wall, which contains exchange and magnetoelastic energy terms. In a first approximation, coercivity can be taken as [17]

$$H^* = \alpha (A(3/2)\lambda_s\mu\xi r)^{1/2}, \tag{8.4}$$

where α is a constant proportional to magnetization, A is the exchange constant, and μ is the shear modulus.

H_C follows a similar behavior with the applied torsional stress as for switching field, and fulfills the mentioned law for $\alpha = 1.8 \times 10^6$ Am J^{-1}, if we assume $A = 6 \times 10^{-12}$ J m^{-1} [17]. All these phenomenological results point out to minimum magnetoelastic anisotropy at $\xi = +2.4$ (2π rad m^{-1}), when internal and applied helical anisotropies are nearly balanced. That minimum for the magnetoelastic anisotropy results in the softest magnetic state, giving rise to a minimum value for the coercivity and affecting also the domain width and the domain wall angle (see Figs. 8.25a,b).

The effect of applied torsion on the surface domain patterns is to change the structure from a maze configuration, with zigzag walls, to a helical one, even for very low ξ values as small as 2.4 (2π rad m^{-1}). Moreover, as the magnitude of the applied torsion is increased, in both clockwise and counterclockwise senses, a change in the values for the domain width as well as the wall angle to the wire axis are clearly observed, with a saturation value of 45° for $\xi \geq 10$ (2π rad m^{-1}) applied in the clockwise sense (and −45° for $\xi < -10$ (2π rad m^{-1}), if counterclockwise torsion is applied). A correlation between coercivity and reduced remanence with the applied torsional strain has been also obtained.

References

1. Honkura Y (2002), *J. Magn. Magn. Mater.*, **249,** 375.

2. Han M, Liang DF, and Deng LJ (2005), *J. Mater. Sci.*, **40,** 5573.

3. Jiles DC (2003), *Acta Mater.*, **51,** 5907.

4. Blanco JM, Zhukov A, and Gonzalez J (1999), *J. Magn. Magn. Mater.*, **196,** 377.

5. Gomez-Polo C, Vazquez M, and Knobel M (2001), *Appl. Phys. Lett.*, **78,** 246.

6. Kim CG, Yoon SS, and Vazquez M (2001), *J. Magn. Magn. Mater.*, **223,** L199.

7. Makhnovskiy DP, Panina LV, and Mapps DJ (2001), *Phys. Rev. B*, **63,** 144424.

8. Betancourt I and Valenzuela R (2003), *Appl. Phys. Lett.*, **83,** 2022.

9. Chizhik A, Gonzalez J, Zhukov A, and Blanco JM (2003), *J. Phys. D*, **36** 419.

10. Zhukov A, Vazquez M, Velazquez J, Garcia C, Valenzuela R, and Ponomarev B (1997), *Mater. Sci. Eng. A*, **226–228,** 753.

11. Kronmuller H (1991), Micromagnetic background of hard magnetic materials, in *Supermagnets, Hard Magnetic Materials* (Kluwer Academic Publishers, Netherlands).

12. Chizhik A, Gonzalez J, Yamasaki J, Zhukov A, and Blanco JM (2004) *J. Appl. Phys.*, **95,** 2933.

13. Sablik MJ and Jiles DC (1999), *Phys. D: Appl. Phys.*, **32,** 1971.

14. Sablik MJ and Jiles DC (1999), *IEEE Trans. Magn.*, **35,** 498.

15. Usov NA (1999), *J. Magn. Magn. Mater.*, **203**, 277.

16. Kabanov U, Zhukov A, Zhukova V, and González J (2005), *Appl. Phys. Lett.*, **87**, 142507.

17. Vázquez M, González J, Blanco JM, Barandiarán JM, Rivero G, and Hernando A (1991), *J. Magn. Magn. Mater.*, **96**, 321.

Chapter 9

Magnetization Reversal in Crossed Magnetic Field

The experiments have been carried out using the transverse magneto-optical Kerr effect in perpendicular circular and axial magnetic field. There were two schemes of the experiments, depending on the combination of magnetic fields:

- transverse Kerr effect at sweeping of circular field $H_{\text{CIR-AC}}$, which was produced by an electric current flowing through the wire I_{AC} ($\mp$ axial bias field $H_{\text{AX-DC}}$);
- transverse Kerr effect at sweeping of axial field $H_{\text{AX-AC}}$ ($\pm$ circular bias field $H_{\text{CIR-DC}}$, which was produced by an electric current flowing through the wire I_{DC}.

Figure 9.1 presents the AC axial surface magnetization curves (50 Hz) with the DC electric current I_{DC} as a parameter. In the absence of I_{DC} (Fig. 9.1a), the symmetrical hysteresis loop should be attributed to the rotation of the magnetization from axial to circular direction and nucleation of circular domain in the outer shell of the wire. In the presence of I_{DC}, a shift of the peaks positions and change of the peaks heights are observed. This evolution (Figs. 9.1b–d) shows how the DC current changes the conditions of the rotation of the magnetization and circular domain nucleation.

Magnetic Microwires: A Magneto-Optical Study
Alexander Chizhik and Julian Gonzalez
Copyright © 2014 Pan Stanford Publishing Pte. Ltd.
ISBN 978-981-4411-25-7 (Hardcover), 978-981-4411-26-4 (eBook)
www.panstanford.com

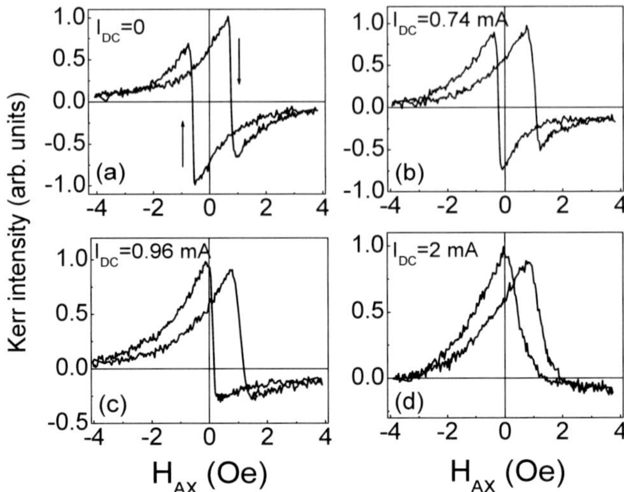

Figure 9.1 Kerr effect dependencies on the AC axial magnetic field in the presence of DC electric current (DC circular magnetic field). Nominal composition $Co_{67}Fe_{3.85}Ni_{1.45}B_{11.5}Si_{14.5}Mo_{1.7}$, metallic nucleus radius R = 11.2 μm, glass coating thickness T = 0.2 μm.

Figure 9.2 presents the AC circular surface magnetization curves with the DC axial magnetic field $H_{AX\text{-}DC}$ as a parameter. In the absence of $H_{AX\text{-}DC}$ (Fig. 9.2a), the shape of hysteresis loop is perfectly rectangular, which is associated with quick enough reversal of circular magnetization. In the presence of $H_{AX\text{-}DC}$, the shift of the hysteresis loop, the decrease in the absolute value of the jump of the magneto-optical signal, and the decrease in coercivity are observed.

The modeling of the axial surface magnetization reversal of glass-covered microwire has been performed taking into account the influence of the DC parameter (I_{DC}) on the nucleation of the circular domains. For conventional microwires, we have performed the modeling of the magnetization reversal in the surface area of the amorphous wire [1]. The formation of circular magnetic multi-domain structure with opposite magnetization directions (bamboo domains) during magnetization reversal was supposed. The model was based on the simultaneous change of the relation of the volumes of the domains with different directions of the circular magnetization and the rotation of the magnetization in the domains. The influence of the DC electric current on the domain volume relation was taken into consideration.

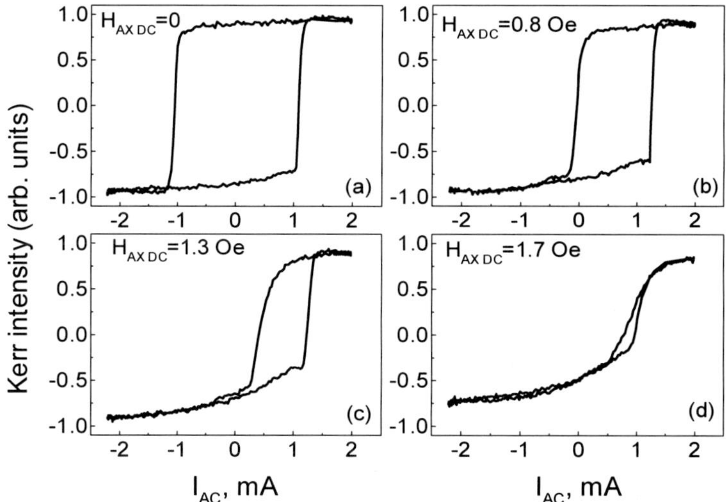

Figure 9.2   Kerr effect dependencies on the AC electric current (AC circular magnetic field) in the presence of DC axial magnetic field.

The perfectly rectangular shape of hysteresis loop presented in Fig. 9.2a is associated with circular magnetic bistability [2]. This means that the multi-domain bamboo-type structure does not occur during the magnetization reversal. Also, it is possible to conclude that the observed shift of the hysteresis loop (Fig. 9.2b) is associated with the influence of the DC parameter on the nucleation process of the circular domains.

Thus, as against the above-mentioned modeling of the magnetization reversal in conventional amorphous wires, here we considered that the circular multi-domain structure does not appear in the studied microwires during magnetization reversal and the large Barkhausen jump between two circular mono-domain states takes place. Also, we supposed that the DC parameter (I_{DC}) favors and delays the nucleation of the circular domains with two different directions of circular magnetization.

Therefore, the magnetization reversal in AC axial magnetic field should be considered in such a way. In the absence of I_{DC} (Fig. 9.1a), when the external axial magnetic field increases from −4 Oe, the monotonic increase of the Kerr signal is observed. This increase is related to the rotation of the magnetization from axial to circular direction in the outer shell of the wire. After that, the sharp jump of the signal with the change of the signal sign takes place. This is related to the nucleation of new circular domain. It is worth mentioning that the sign of the nucleated circular domain is associated with the sign of the external magnetic field. In the presence of DC current, a shift of the peaks positions and change of the peaks values are observed. This evolution shows how the DC current changes the conditions of the circular domain nucleation. The DC electric current of a determined direction assists nucleation of circular domain of a determined direction that is reflected in the shift of the peaks. When the electric current is high enough (Fig. 9.1d), only one domain exists and only the rotation of magnetization is observed.

The transverse Kerr effect reflects the change of the part of the magnetization, which is perpendicular to the field-light plane (perpendicular magnetization). Thus, in our analysis, we consider the field dependencies of the perpendicular component of the magnetization in order to compare calculation and experiment.

Modeling our experimental results, we consider that the magnetization reversal process consists of rotation of magnetization and nucleation of magnetic domain. When the axial magnetic field changes from "–" to "+", the rotation of magnetization ($M_{1\text{perp}}(H)$) appears as shown in Fig. 9.3a. The shape of this curve has been chosen for modeling to be closed to the experimental curve (Fig. 9.1d, right part of the hysteresis loop), which describes the rotation of magnetization in one surface domain, when the axial magnetic field changes from "–" to "+". When the magnetization is closed to circular direction, the nucleation of the domain with opposite direction of circular magnetization appears. The field dependencies of the annihilation ($V_1(H)$) and the nucleation of circular domains ($V_2(H)$), which was used in the modeling, are presented in Figs. 9.3c,d. The rotation of magnetization toward the axial direction takes place after the nucleation process. Accordingly, the expression can be written as follows:

$$M_{\text{perp}} \rightarrow (H) = \{M_{1\text{perp}}(H) \times V_1(H)\} - \{M_{1\text{perp}}(H) \times V_2(H)\}, \quad (9.1)$$

where $M_{\text{perp}} \rightarrow (H)$ is the field dependence of the perpendicular magnetization in the outer shell of glass-covered microwires when the axial magnetic field changes from "–" to "+". The "–" before the second member of the equation reflects the fact, that the sign of the circular projection of the magnetization has been changed after the nucleation.

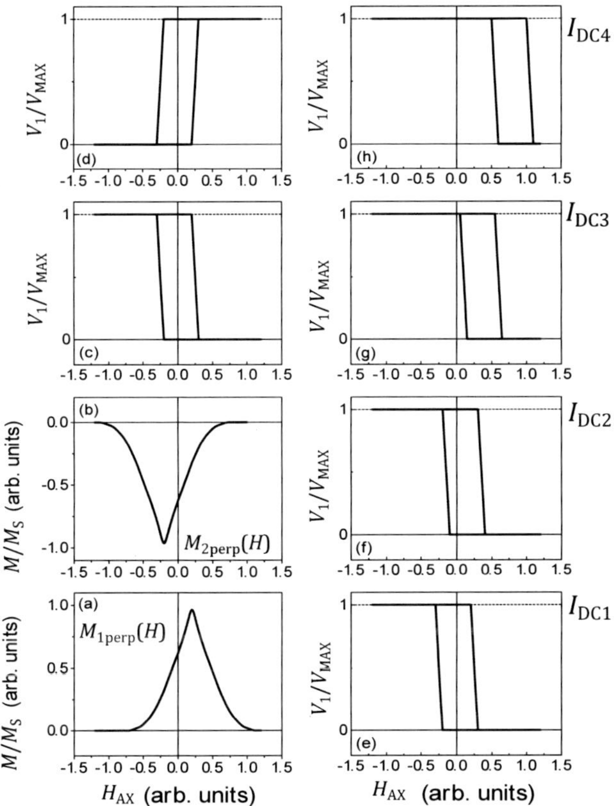

Figure 9.3 Axial magnetic field dependencies of $M_{1perp}(H)$ (a), $M_{1perp}(H)$ (b), $V_1(H)$ (c, e–h), and $V_2(H)$ (d), which was used in modeling. $I_{DC1} < I_{DC2} < I_{DC3} < I_{DC1}$, $I_{DC1} = 0$.

When the axial magnetic field changes from "+" to "–", the following expression can be written as

$$M_{perp}\!\leftarrow\!(H) = \{M_{2perp}(H) \times V_2(H)\} - \{M_{2perp}(H) \times V_1(H)\}, \qquad (9.2)$$

where $M_{perp}\!\leftarrow\!(H)$ is the field dependence of the perpendicular magnetization in the outer shell of glass-covered microwires when the axial magnetic field changes from "+" to "–". $M_{2perp}(H)$ is the dependence, which described the rotation of magnetization when the axial magnetic field changes from "+" to "–" (Fig. 9.3b). The calculated dependencies of $M_{perp}\rightarrow(H)$ and $M_{perp}\leftarrow(H)$ are presented in Fig. 9.4.

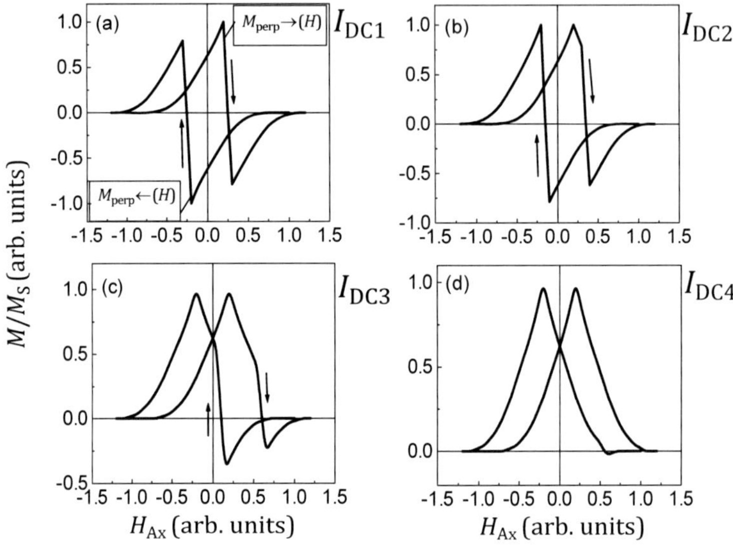

Figure 9.4 Calculated dependencies of $M_{perp}\rightarrow(H)$ and $M_{perp}\leftarrow(H)$. $I_{DC1} < I_{DC2} < I_{DC3} < I_{DC1}$, $I_{DC1} = 0$.

In the second stage of the modeling, we took into consideration the influence of the DC electric current on the nucleation process. Taking into account that the DC electric current of a determined direction assists the nucleation of the circular domain of a determined direction and inhibits the nucleation of the domain of the opposite direction, the shift of the $V_1(H)$ and $V_2(H)$ dependencies along the H_{AX} axis should be supposed. The $V_1(H)$ dependencies, which were used in our modeling for the case of the presence of DC electric current, are presented in Figs. 9.3e–h. The method of the $M_{perp}\leftarrow(H)$ and $M_{perp}\leftarrow(H)$ calculation was the same as in the first stage of the modeling. The modeling was performed under the conditions of $I_{DC1} < I_{DC2} < I_{DC3} < I_{DC1}$, $I_{DC1} = 0$ and $V_2(H) = 1 - V_1(H)$. The results of the modeling for the case of the application of DC electric current are presented in Figs. 9.4b–d.

The good agreement of the experimental results (Fig. 9.1) and the results of the modeling (Fig. 9.4) have been obtained. It allows us to conclude that the main assumption of the model is adequate. The magnetization reversal process in the outer shell of the studied glass-covered microwires is mainly determined by the rotation of the magnetization and the circular domain nucleation. The multi-domain bamboo structure does not appear. The absence of the bamboo-like domain structure in the case of glass-coated microwires could be attributed to the enhanced magnetoelastic energy due to the additional internal stresses appearing from the glass coating. In this case, the energy of bamboo-like domain walls also increases. We assume that such increasing of the domain wall energy might result in the disappearance of such bamboo-like domains.

The nucleation of the circular domains is very sensitive to the DC electric current, which produces the circular magnetic field. Some difference between the experimental and calculated hysteresis loops could be related to the fact that the DC electric current could cause not only the shift of the $V_1(H)$ and $V_2(H)$ dependencies but also the change of the shape of these loops.

References

1. Chizhik A, Zhukov A, Blanco JM, and Gonzalez J (2001), *Phys. B*, **299**, 314.

2. Chizhik A, Gonzalez J, Zhukov A, and Blanco JM (2003), *J. Phys. D*, **36**, 419.

Chapter 10

Visualization of Barkhausen Jump

10.1　Introduction

The direct experimental observation of the fundamental effect is remarkable just as a special event in modern physics—especially when such an effect is a fundamental phenomenon that forms the basis of a special direction in science. In the pioneering works of Sixtus and Tonks [1, 2], it was demonstrated that the magnetization reversal in magnetic wires takes place through a giant Barkhausen jump. The last work of this series of investigations was published in 1938, but in their interpretation the mentioned authors did not use the Landau–Lifshitz domain model published in 1938. A new stage in the investigation of the magnetic wire came in the 1980s, when the effect of the giant Barkhausen jump was detected in as-quenched amorphous wires as a sharp voltage pulse induced in a small coil placed across the wire or as voltage generated by the wire itself because of the Matteucci effect [3]. In spite of the fact that observation of surface magnetic domains in amorphous wires has shown that the giant Barkhausen jump is related to a transformation of the domain structure, direct observation of the giant Barkhausen jump in the predicted form of domain wall motion has not been performed during the long time.

Magnetic Microwires: A Magneto-Optical Study
Alexander Chizhik and Julian Gonzalez
Copyright © 2014 Pan Stanford Publishing Pte. Ltd.
ISBN 978-981-4411-25-7 (Hardcover), 978-981-4411-26-4 (eBook)
www.panstanford.com

10.2 Experiment

There were two configurations of the MOKE microscopy experiments: In the first, the evolution of magnetic domain structure was determined from comparison of remanent domain structure images in the zero field and after application of electric current pulses with a constant time duration of pulse and increasing amplitude I. In the second case, the domain structure transformation was obtained after applying I with a constant value of amplitude and increasing time duration pulses t.

During our study, we performed MOKE magnetometer experiments and, later, investigation of the transformation of magnetic domain structure using the MOKE microscope. The dependencies of the Kerr intensity on the electric current (circular magnetic field) are presented in Fig. 10.1. These dependencies reflect the surface magnetization reversal process [4]. During experiments, the DC axial bias field (H_{AX}) was applied along the microwire axis. The jumps of the circular magnetization related to Barkhausen jumps can be observed in Fig. 10.1. These jumps are related to the mentioned effect of circular magnetic bi-stability. It was found that applying an axial magnetic field (Fig. 10.1b) transforms the hysteresis curve in such a way that the jump of the circular magnetization becomes sharper. The observed change of the hysteresis loop is related to the transformation of the helical magnetic structure [5] induced by the axial magnetic field. In the presence of an axial field of some special value, the jump occurs between two circularly magnetized states. For an axial field with a value of 0.27 Oe, the sharpness of the jump is maximal.

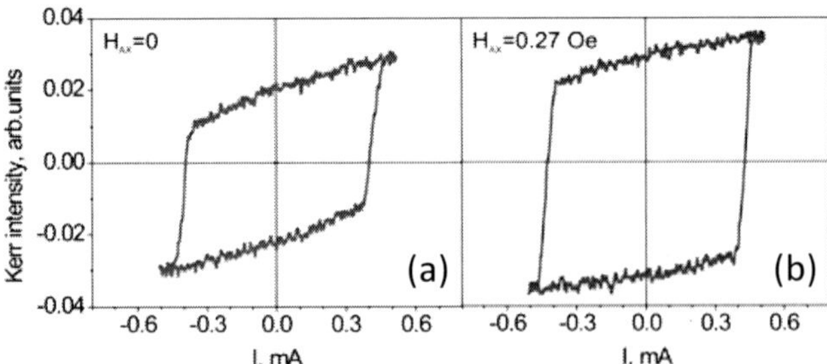

Figure 10.1 Hysteresis loops measured by MOKE as a function of the electric current (circular magnetic field) for different axial magnetic field: (a) $H_{AX} = 0$, (b) $H_{AX} = 0.27$ Oe.

Thus, we have two processes with different velocities of magnetization reversal. We assume that this difference in velocity reflects the difference in the process of domain nucleation and domain walls propagation. To verify this supposition and to elucidate the details of the Barkhausen jumps, we performed the MOKE microscope experiments. The results of these experiments are presented in Figs. 10.2 and 10.3.

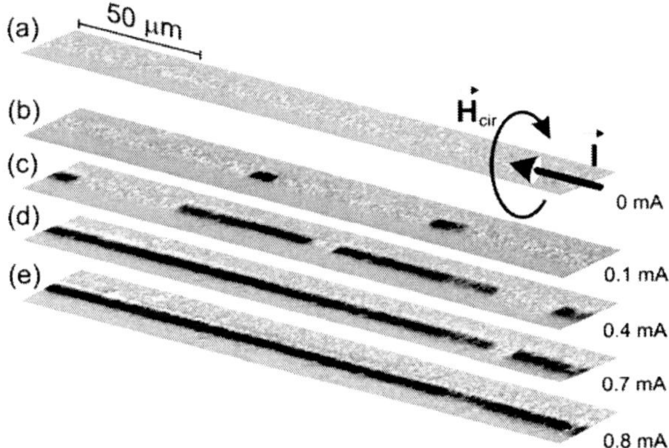

Figure 10.2 Images of circular magnetic domain structure transformation in the presence of electric current without axial magnetic field: (a) $I = 0$, (b) $I = 0.1$ mA, (c) $I = 0.4$ mA, (d) $I = 0.7$ mA, (e) $I = 0.8$ mA.

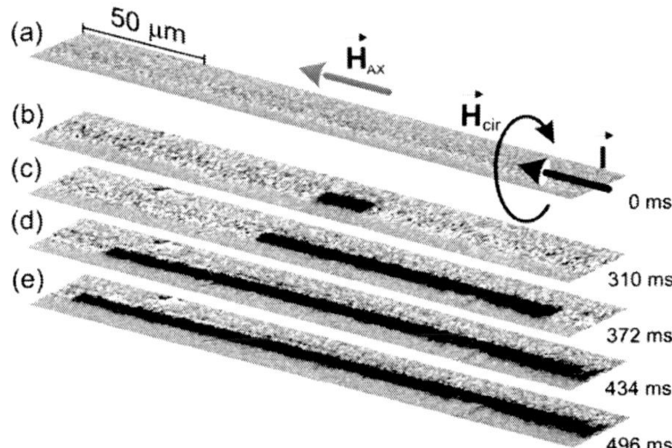

Figure 10.3 Images of circular magnetic domain, in the electric current $I = 0.38$ mA with $H_{AX} = 0.27$ Oe, recorded at different times: (a) $t = 0$, (b) $t = 310$ ms, (c) $t = 372$ ms, (d) $t = 434$ ms, (e) $t = 496$ ms.

We performed the experiments with increasing amplitude of the electric current (circular magnetic field) for the case of $H_{AX} = 0$ as well as the experiments with increasing time duration of the circular magnetic field with constant amplitude, for the case of $H_{AX} = 0.27$ Oe. A second experimental configuration was chosen for the case of $H_{AX} \neq 0$, taking into account the very sharp jump of circular magnetization, which takes place in the presence of the axial magnetic field. The circular field range of the existence of a giant Barkhausen jump is quite short; therefore, time scanning was applied to observe the details of the magnetization reversal process.

When $H_{AX} = 0$, the successive increase of the circular magnetic field induces the nucleation of several of the circular magnetic domains in the surface of the microwire (Fig. 10.2), followed by the propagation of domain walls along the microwire. Three different stages can be highlighted. First, the nucleation of two domains can be observed in the field of vision of the microscope. Second, parts of two other domains, which come into the field of vision, can be observed along with the propagation of the two nucleated domains. Finally, the joint motion of a number of domain walls is finished by the collapse of the disadvantageous circular domain (Fig. 10.2e).

Figure 10.3 shows the surface magnetization reversal for the case of $H_{AX} \neq 0$. The successive increase of the time duration of the constant circular magnetic field induces the nucleation of a solitary circular domain followed by the propagation of two domain walls moving into two opposite directions along the microwire. This scenario could be considered as a classic realization of the giant Barkhausen jump.

Thus, we can see that the axial bias magnetic field affects not only the surface helical magnetic structure but also the surface giant Barkhausen jump. In particular, the axial magnetic field changes the mechanism of formation of surface domain structures. When $H_{AX} = 0$, the process of multi-domain nucleation takes place. This process depends basically on the local distribution of the density of nucleation centers on the surface of the microwire [6]. The increase of the axial field suppresses mainly the nucleation process of the surface circular domains. Consequently, the quickest version of the giant Barkhausen jump is determined by the propagation of the domain wall (or of two walls), which follows the nucleation of one circular domain. In some sense, the influence of the axial field on the surface helical structure is strongly related to its influence on the nucleation process. The angle of the helicality (the direction of the circular magnetization in the surface area of the microwire), which can be controlled by the axial magnetic field, determines the probability of the nucleation of the reversed domain and in this way establishes the relative priority of the domain nuclcation and the domain walls propagation.

We studied a series of microwire samples from different parts of the bobbin to verify the reproducibility of the obtained results along the length of the microwire. The formation of a surface multi-domain structure was observed in the whole studied series. However, from the series of microwires within the 0.79–0.98 range of ρ, the formation of a multi-domain structure was observed only in the microwire with $\rho = 0.79$. This confirms that the formation of multi-domain structure depends strictly on the internal stress distribution, which is determined in particular by the glass thickness of the microwire.

Using an axial magnetic field as a unique external parameter, we have determined that the surface magnetization reversal is realized in the form of a sharp giant Barkhausen jump in magnetic microwires. We have demonstrated that this jump is characterized by the quick motion of the solitary domain wall, which overpasses long distances of approximately hundreds of micrometers. We have shown that the extensive motion of the solitary domain wall really exists in magnetic wires and that this is the determinative constituent of the surface giant Barkhausen jump.

Also, as a result of the MOKE magnetometry experiments, the series of the Kerr intensity dependencies on the electric current I (circular magnetic field) in the presence of DC bias H_{AX} has been obtained (see top panel in Fig. 10.4). The strong transformation of hysteresis curves induced by H_{AX} has been found. The jump of the circular transversal magnetization becomes smoother and smaller with axial magnetic field. Also, the values of the normalized remanence magnetization $**M_R$ and the switching field H_{SW} (see Fig. 10.4) decrease with H_{AX} increase. The DC axial field produces the pronounced shift of the hysteresis loop along the horizontal axis. The direction of the shift depends on the sign of the axial field. When the axial magnetic field reached the special value, the hysteresis was not observed and the magnetization reversal took place as coherent rotation of magnetization. The obtained results are sketched as the dependencies of the normalized M_R and H_{SW}^* (normalized to maximum value) on DC axial field and are presented in Figs. 10.4 and 10.6. These dependencies have a maximum at the value about 0.27 Oe.

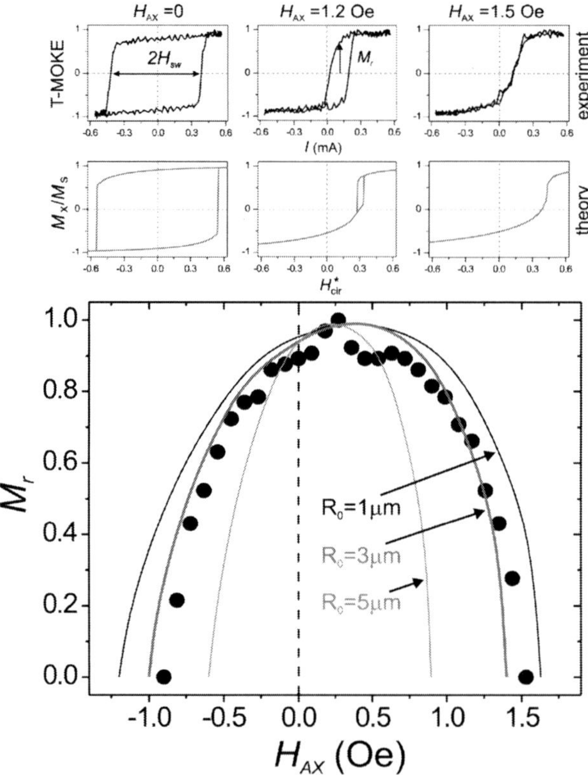

Figure 10.4 Top panel: the measured hysteresis loops and the loops– M_X/M_S vs. H_{CIR}* calculated from Eq. (10.1) for H_{AX} = 0, 1.2 Oe and 1.5 Oe. Dependence of normalized remanent circular magnetization component M_r on the external axial magnetic field H_{AX} (points). Lines are calculated within the proposed model, assuming the anisotropy constants: h_t = 2.9 × 10⁻⁴, h_a = 4.8 × 10⁻⁴, α = 20° and the radius inner core R_0 = 1 μm, 3 μm, 5 μm.

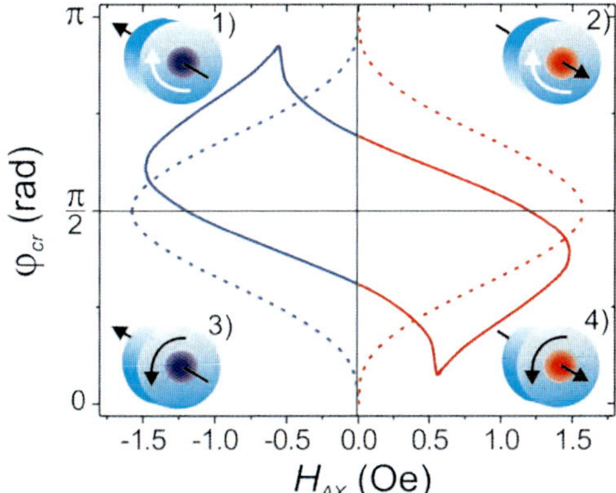

Figure 10.5 The critical angle φ_{cr} as functions of the axial applied magnetic field H_{AX} for $\alpha = 0$ (dashed line) and $\alpha = 20°$ (solid line). There are one-to-one correspondences between the blue (red) branches in the upper and bottom figures. Four magnetization states (1–4) corresponding to branches of $\varphi_{cr}(H_{AX})$ where the red and blue colors indicate the two directions of the axial magnetization (forward and backward).

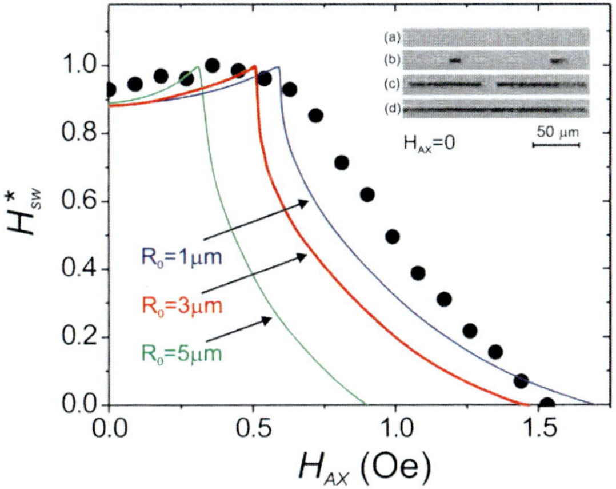

Figure 10.6 Switching field H_{SW}^* on function the applied axial magnetic field H_{AX}. The points are the measured while the lines represent the H_{SW}^* calculated from Eq. (10.1) for radius inner core $R_0 = 1$ μm, 3 μm, 5 μm. The inset shows the magnetic domain structures without H_{AX} for different current: 0 (a), 0.1 mA (b), 0.4 mA (c), 0.8 mA (d). The wire was initially saturated (gray area) by the circular magnetic field $H_{CIR} > 0$. The black areas illustrated to the change of the structure induced by $H_{CIR} < 0$ pulses.

10.3 Calculation

The obtained results could be interpreted in the frameworks of the phenomenological model, which is supposed the external circular and axial field induced transformation of the surface magnetic structure. We consider that originally without axial magnetic field, the studied microwire has a helical magnetic structure in the surface area. The application of the axial magnetic field causes the transformation the surface magnetic structure and in particular, the change of the angle of the helicality (the direction of the circular magnetization in the surface area of the microwire). When the axial field is about 0.27 Oe the surface magnetization is directed strictly circularly. For this value of axial field the surface magnetization reversal occurs as the jump between two circular states and the jump of the circular magnetization has a maximum value. The angle of the helicality could be controlled by the axial magnetic field H_{AX}. When the axial field is high enough the magnetization with axial direction is stable that causes the disappearance of the circular hysteresis loop.

We assume that the magnetic structure of the studied microwire consists of a soft inner core with axial anisotropy and the outer shell in which the helical magnetic anisotropy dominates. In other words, a wire consists of two different areas—the core and the shell—with different magnetic anisotropies: axial and helical, accordingly. The core radius is changeable and it depends on magnetic anisotropy distribution inside the wire. To determine the radius of the inner core, we have performed the fluxmetric experiments [7] when the axial hysteresis loop has been measured as a dependence of the axial magnetic field. Following [8], the value of the inner core radius was determined about 3 μm.

Let us describe magnetization states of the wire assuming that the magnetization vector has two components: circular and axial. We denote as φ the angle between the magnetization and the circular field (produced by current). As mentioned above, the inner core is assumed to have only axial magnetic anisotropy, and from hereinafter we denote its radius as R_0. The total energy of such a domain configuration is the sum of the anisotropy, Zeeman and exchange energies:

$$E = 2\pi^2 M_S^2 D((R^2 - R_0^2)h_t \sin^2(\phi + \alpha) + R_0^2 h_a \cos^2(\phi + \alpha))$$

$$-(2R^2(h_{CIR}\cos\phi + h_{AX}\sin\phi) + 2l_{ex}^2\cos^2\phi \ln(R/l_{ex})), \qquad (10.1)$$

where $l_{ex} = (A/2\pi M_S^2)^{1/2}$ is the exchange length, A is the exchange constant, and D is the domain length along the wire; α is the angle between the helical anisotropy easy axis with the transversal direction, $h_a = H_{Aa}/4\pi M_S$ and $h_t = H_{At}/4\pi M_S$ are normalized axial and tangential (circular) the anisotropy fields (where $H_{At} = 2K_t/M_s$ and $H_{Aa} = 2K_a/M_S$ are the anisotropy fields), $h_{CIR} = H_{CIR}/4\pi M_S$ is the scaled circular component of the applied field (due to current) and $h_{AX} = H_{AX}/4\pi M_S$. In Eq. (10.1), the last term represents the azimuthal exchange energy in the form derived in [9]. Note that in Eq. (10.1) the magnetostatic energy contribution is zero because of the closed magnetic flux structure, and we neglect the magnetic poles on the wire edges (a long wire). We now minimize Eq. (10.1) with respect to angle φ and calculate magnetization hysteresis loops—M_X/M_S as a function of $H_{CIR}^* = H_{CIR}/H_{At}$ shown in Fig. 10.4 (top panel). A good agreement between the experimental and calculated magnetic hysteresis loops is seen from Fig. 10.4. In Fig. 10.4, for small applied axial fields $H_{AX} = 0$ and $H_{AX} = 1.2$ Oe, the shown curves demonstrate the jump-like transitions between helical states of different chiralities. On the circular field scale, the jump points—the switching fields (H_{SW})—are defined by the conditions: $dE/d\varphi = 0$ and $d^2E/d\varphi^2 = 0$. The solution of these two equations gives us the switching circular field and critical angle of the magnetization φ_{cr}. In Fig. 10.5, we plot the critical angle as a function of the applied axial field for $\alpha = 0$ and $\alpha = 20°$. For each α, there are the four critical states corresponding to different chiralities (left-handed and right-handed) and axial orientation (forward and backward) of the wire magnetization (see Fig. 10.5). The actual appearance of one of these four states depends on magnetic prehistory of a wire. On changing H_{AX}, the critical angle changes between 0 and π, and for $\alpha = 0$ when $H_{AX} = 0$, the shown states with $\varphi_{cr} = 0$ or $\varphi_{cr} = \pi$ correspond to the pure symmetrical circular magnetization states. For $\alpha^1 \neq 0$, the magnetization states are asymmetrical and describe four helical magnetization states of different chiralities. Thus, in low applied axial magnetic field, a jump-like transition between helical states of different chiralities takes place when the circular field (current) reaches the switching field value.

Equation 10.1 allows us to calculate, using the values of the magnetic anisotropy constants from [10], both circular magnetization in the remanence state (M_R) and the normalized switching field

$H_{SW}{}^*$ as functions of the axial magnetic field for different radii of the inner core R_0. These results are shown in Figs. 10.4 and 10.6, correspondingly. In Fig. 10.4 the calculated M_R are compared with the experimental data obtained for the wire with inner core radius $R_0 = 3$ μm and $R = 11.2$ μm; there is a good agreement between the experimental and the calculated data. The $M_R(H_{AX})$ curves calculated for $R_0 = 1$ μm and $R_0 = 5$ μm predict the increase and decrease of the remanent magnetization, accordingly. In Fig. 10.6, we demonstrate the calculated and experimental curves: H_{SW}^* vs. H_{AX} for the wire with three different core radii: $R_0 = 1$ μm, 3 μm, and 5 μm. Here the coincidence between the experimental and calculated curves is only satisfactory in the vicinities of the starting and ending points of the field range change, $0 < H_{AX} < H_{AX0}$. This discrepancy is explained by a domain structure that exists namely in this field range. Indeed, the magnetization states are barely circular or axial just at $H_{AX} = 0$ so that they are monodomain states. In the moderate axial field, helical magnetic domains with different chiralities appear (see the inset to Fig. 10.5 as well as the experimental data shown in the bottom panel of Fig. 10.4). In order to switch the existing domain structure, an enhanced circular applied field is required because of an energy barrier related to the domain walls. Indeed, in a wire, magnetic domains of opposite chirality are separated by 180° domain walls. On increasing the current (circular field) the domains magnetized antiparallel to H_{CIR} are squeezed and finally collapse. However, before the collapse, the approaching domain walls could constitute 360° walls, and namely they lift the measured switching fields. This effect is similar to that known from physics of bubble materials: In low-coercivity garnet films, the existence of such domain walls leads to a significant increase of the saturation field—the field of the transition to the monodomain state [11]. A similar effect of the enhanced field stability of 360° domain walls was recently observed in ion beam irradiated lines of ultrathin Pt/Co(1.4 nm)/Pt films. Note that in a wire with no radial anisotropy, there is no the magnetostatics reason for the domain appearance. In such wires, the formation of irregular circularly magnetized domains was explained by heterogeneous nucleation processes taking place on applying a superposition of axial and circular fields. These domains do not correspond to the energy minimum with respect to their period, and therefore such domains should be treated as metastable; this is a signature of magnetism of low dimensional systems. Other examples of metastable domains can be found in Ref. [12].

The existence of the four magnetization states with different chiralities and axial directions is proven by experiment and theory. We proposed a phenomenological model that describes magnetization states and their current-induced switching in wires with variable inner core radius. In low applied axial fields, on increasing the current (circular field), the initial left-handed chirality circular magnetization states undergoes the following phase sequence: a continuous transformation to a helical structure, then a jump to a helical state with the opposite chirality, and finally, gradual magnetization rotation to the right-handed chirality circular state. An applied axial magnetic field smoothes the jump between the states with the opposite chiralities and makes magnetization states asymmetrical. We highlighted the role of the inner core in the formation of the magnetization states and magnetization reversal in such wires. This opens new directions of experimental study of wire within wide range of thickness magnetization configurations affected by external factors enabling a redistribution of internal stress and therefore wire anisotropy, e.g., applied stress and temperature changes.

References

1. Sixtus KJ and Tonks L (1931), *Phys. Rev.*, **37**, 930.

2. Sixtus KJ and Tonks L (1932), *Phys. Rev.*, **39**, 357.

3. Mitrat A and Vazquez V (1990), *J. Phys. D: Appl. Phys.*, **23**, 228.

4. Chizhik A, Gonzalez J, Zhukov A, and Blanco JM (2003), *Appl. Phys. Lett.*, **82**, 610.

5. Chizhik A, Garcia C, Zhukov A, Gonzalez J, Gawronski P, Kulakowski K, and Blanco JM (2008), *J. Appl. Phys.*, **103**, 07E742.

6. Ipatov M, Usov NA, Zhukov A, and Gonzalez J, (2008), *Phys. B*, **403**, 379.

7. Zhukov A, Vázquez M, Velásquez J, Chiriac H, and Larin V, (1995), *J. Magn. Magn. Mater.*, **151**, 132.

8. Severino AM, Gomez-Polo C, Marin P, and Vázquez M, (1992), *J. Magn. Magn. Mater.*, **103**, 117.

9. Hoffmann H and Steinbauera F, (2002), *J. Appl. Phys.,* **92**, 5463.

10. Graus L, Vázquez M, Infante G, Badini-Confalonieri G, and Torrejón J, (2009), *Appl. Phys. Lett.,* **94**, 062505.

11. Gemperle R, Murtinova L, and Kaczer J, (1985), *Acta Phys. Slov.,* **35**, 216.

12. Zablotskii V, Stefanowicz W, and Maziewski A, (2007), *J. Appl. Phys.,* **101**, 113904.

Chapter 11

Magnetization Reversal in Glass-Covered Nano-Wires of Cylindrical Shape

Following the tendency of the miniaturization of active elements for magnetic sensors, the MOKE investigation of the magnetization reversal has been performed in Fe-rich nanometric amorphous wires (nominal composition $Fe_{72.75}Co_{2.25}B_{15}Si_{10}$).

First, the arrays of glass-covered microwires (radius of metallic nucleus 500 nm) have been studied by magnetic. The result of PPMS magnetic (four-nano-wire array) experiments is presented in Fig. 11.1. The clear jumps of magnetization could be observed in volume. These jumps are related to the giant Barkhausen jumps associated with the magnetization reversal in single microwire as a consequence of the interaction between the microwires. Based on the obtained results, we can conclude that the magnetic behavior of the glass-covered wires with such extremely tiny diameter remains magnetically bistable and could be considered in the frame of the core–shell model.

Second, the arrays of glass-covered nano-wires (radius of metallic nucleus 500 nm) have been studied by magneto-optical techniques (Fig. 11.2). During the experiment, the array of 10 wires has been rotated in the plane perpendicular to the plane of the light. The results of longitudinal experiments are presented in Figs. 11.3 and 11.4.

Magnetic Microwires: A Magneto-Optical Study
Alexander Chizhik and Julian Gonzalez
Copyright © 2014 Pan Stanford Publishing Pte. Ltd.
ISBN 978-981-4411-25-7 (Hardcover), 978-981-4411-26-4 (eBook)
www.panstanford.com

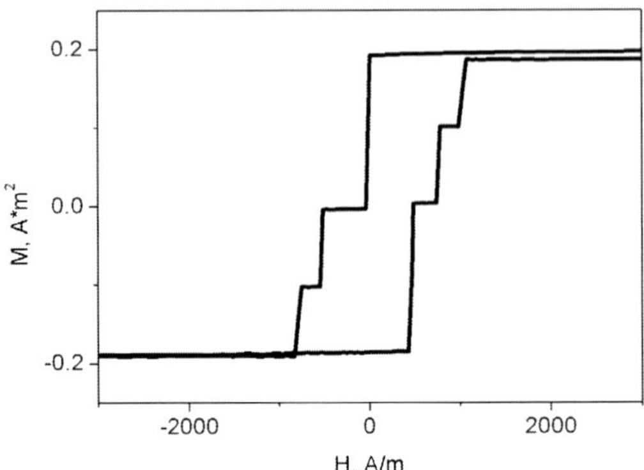

Figure 11.1 Magnetization reversal curves obtained by magnetic techniques.

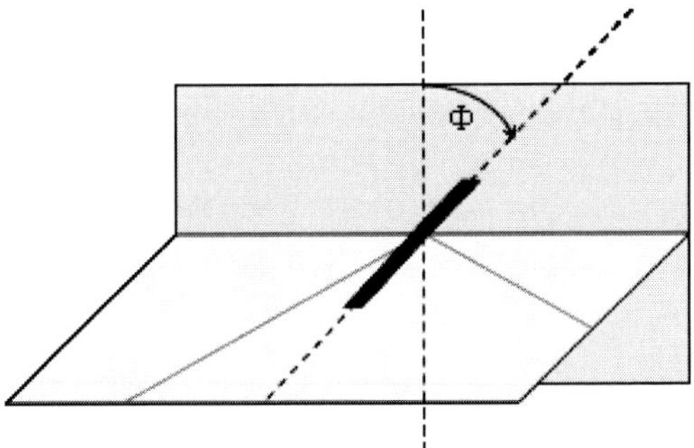

Figure 11.2 Schematic picture of MOKE experiment.

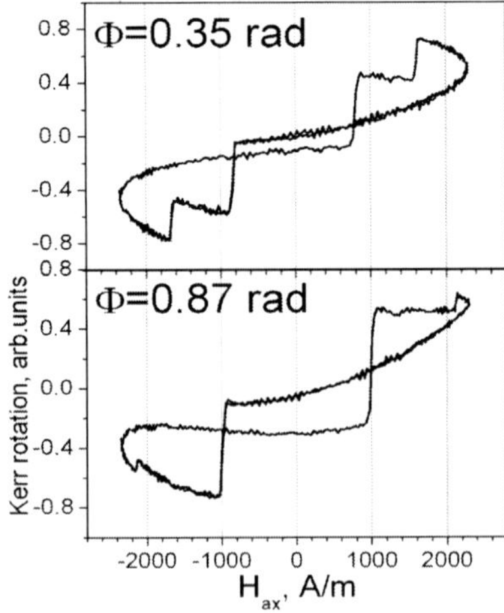

Figure 11.3 MOKE hysteresis loops for two different value of the angle Φ.

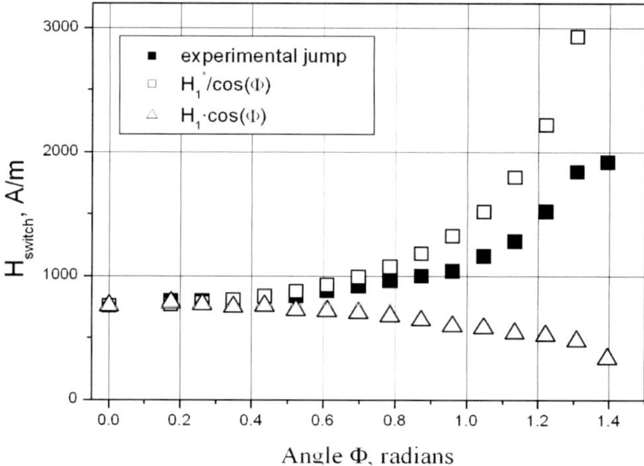

Figure 11.4 Dependencies of the switching fields H_1 on angle Φ.

The clear jumps of magnetization could be observed in surface hysteresis loops (Fig. 11.3). These jumps are related also, as for the case of volume loops, to the giant Barkhausen jumps associated with the magnetization reversal in single microwire as a consequence of the interaction between the microwires.

Although the array contains 10 wires, there are only two jumps in the MOKE hysteresis loop. This means that in these experiments the collective jumps of some number of wires take place. For example, six microwires change the direction of the magnetization during the first jump and then four microwires change the direction of the magnetization during the second jump. This behavior could be explained by the very narrow distribution of the switching field in the studied microwires.

Figure 11.4 presents the angle Φ dependence of the switching fields H_1. As it is possible to see, the Barkhausen jumps take place up to the angle Φ of about 80°. This means that the bistability effect exists for the almost perpendicular direction of the microwire to the external magnetic field. Analyzing the angle dependence of the first jump, we have constructed the dependence $H_1^*/\cos(\Phi)$ (white quadratic points), where H_1^* is the value of the switching field for the angle $\Phi = 0$. Also, we have constructed the dependence $H_1 \cos(\Phi)$ (white triangle points).

For the simplest case, when only the projection of the external magnetic field on the wire axis causes the magnetization reversal, we could obtain the following results: Black and white points should coincide and the blue points should lie on the linear horizontal dependence. It is not observed in our experiments. Basically, the switching field H_1 decreases with the angle that confirms the fundamental role of the axial magnetic field in the magnetization reversal. However, in the real experiments, the same value of the switching field could be reached for the higher value of the angle Φ. The difference between the experimental points and the result of the calculation testifies that axial magnetic field also affects the process of the nucleation of the magnetic domains: inclined magnetic field induces the nucleation of the reversed magnetic domain. Therefore, we have two contrary effects: On the one side, it is needed to apply the higher magnetic field to remagnetize the wire for inclined configuration, but on the other side, inclined magnetic field provokes the domain nucleation process.

Also, we have performed MOKE study of the series of the nano-wires with different values of geometric ration: sample No. 1 ρ = 0.04, metallic nucleus radius r = 400 nm, D = 19 μm; sample No. 2 ρ = 0.067, r = 700 nm, D = 21 μm; sample No. 3 ρ = 0.085, r = 1000 nm, D = 21 μm. A tiny Cu wire was attached to the sample end in order to apply the axial tensile stress. The torsion stress has been applied during the experiments too.

First, it was confirmed that for single nano-wire, the surface hysteresis loop has a rectangular shape related to magnetic bistability effect (Fig. 11.5).

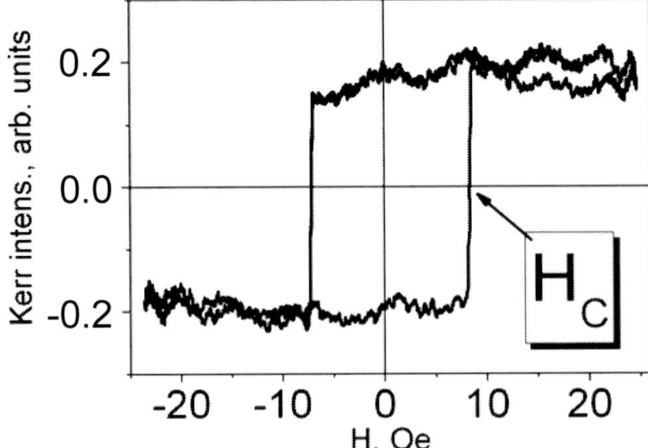

Figure 11.5   MOKE dependence on axial magnetic field (sample No. 3).

Figure 11.6 demonstrates the coercive field (H_C) growth with decreasing of the geometric ratio ρ. The highest value of the surface coercive field is observed for the extremely small value of the ratio ρ. Such rising of the switching field has been attributed to the increase of the strength of internal stresses as the glass coating thickness is increased.

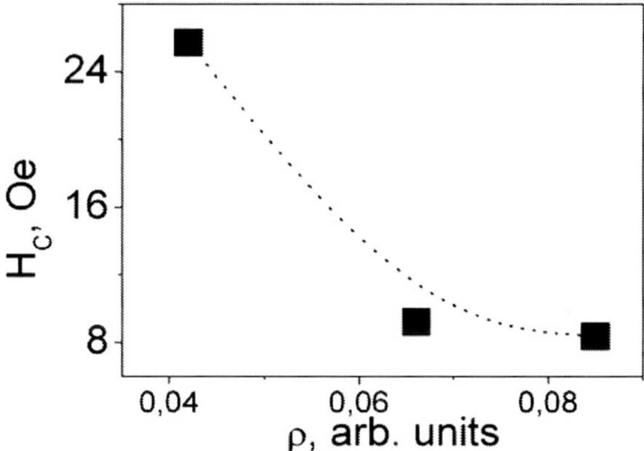

Figure 11.6 Dependence of surface coercive field on geometric ration ρ.

The experimental H_C dependence on the tensile stress has been plotted as a function of the square root of applied stress σ (Fig. 11.7). A good fitting of the experimental points by the linear dependence takes place. The surface bistability effect is related to formation of surface domain wall. The coercive field in the surface like in the volume of the wire is proportional to the energy required to form the domain wall γ involved in the bistable process.

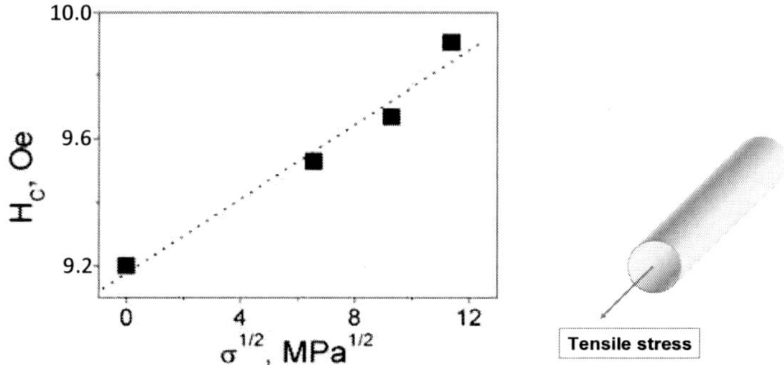

Figure 11.7 Tensile stress dependence of coercive field for sample No. 2.

The surface coercive field is related to the magneto-elastic anisotropy as given by [1]

$$H_C \sim \gamma \sim [(3/2)\,A\lambda_S(\sigma + \sigma_r)]^{1/2},\qquad (11.1)$$

where A is the exchange energy constant, λ_S is saturation magnetostriction constant, σ is applied tensile stress, and σ_r is the internal tensile stress. As it is possible to see, the coercive field must be proportional to $\sigma^{1/2}$ for the applied stress σ larger than the internal stress that is observed in the performed experiments.

The experimental H_C dependence on the torsion stress also has been plotted as a function of the square root of applied stress τ (Fig. 11.8).

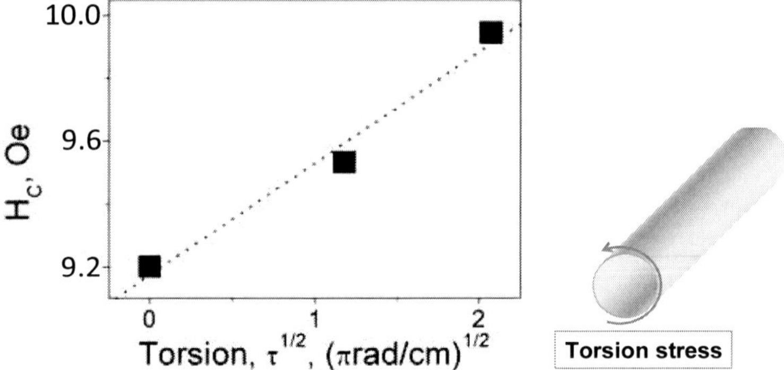

Figure 11.8 Torsion stress dependence of coercive field for sample No. 2.

Analysis of the obtained results has been performed following reference [2], in which the appearance of the magneto-elastic anisotropy coming from the applied torsion stress is supposed. In this case, the dependence of the coercive field on the torsion stress could be presented as

$$H_C \sim \gamma \sim [(3/2)\,A\lambda_S\tau]^{1/2}\qquad (11.2)$$

A good fitting of the experimental points by the linear dependence also observed for the torsion stress that permits us to suppose the existence of the "inner core–outer shell" magnetic configuration with radial and surface closure domains in the surface of the studied nano-wires.

In the extremely thin Fe-rich glass-covered nanometric wires, the magnetic bistable behavior is observed like in the glass-covered wires of micro-scale. The performed analysis of the tensile and torsion stresses transformation of surface hysteresis loop demonstrates that about one order decrease of the wire scale does not abolish the basic effects observed earlier in thicker wires. It permits us to reduce considerably the size of basic elements of magnetic sensors and make the next step in the way of sensor miniaturization.

References

1. Kraus L, Kane SN, Vazquez M, Rivero G, Fraga E, Hernando A (1994), *J. Appl. Phys.*, **75,** 6952.

2. Raposo V, Gallego JM, Vazquez M (2002), *J. Magn. Magn. Mater.*, **242–245,** 1435.

Chapter 12

Magnetic Domain Wall Dynamics in Co-Rich Glass-Covered Microwires

12.1 Introduction

The investigation of the DW motion in magnetic microwires is an interesting topic in the scientific research because of its great perspectives of the technological application. The dynamics of the DW in microwires with positive magnetostriction was studied widely during the last years using traditional Sixtus-Tonks method. Generally, in traditional Sixtus-Tonks method, the pulse of axial magnetic field was used to drive the DW. The dynamics of the surface DW in microwires with negative magnetostriction was not studied ever before although the dynamics of the surface circular domain walls is the key process of the GMI effect. Earlier, we have demonstrated the existence of surface giant Barkhausen jump which is characterized by the quick motion of the solitary circular domain wall in microwires with negative magnetostriction. This DW could overpass enough long distances. It divides the circularly magnetized domains in the surface of the microwires. Also it was found that the circular magnetic field could be used to drive the surface circular DW.

Magnetic Microwires: A Magneto-Optical Study
Alexander Chizhik and Julian Gonzalez
Copyright © 2014 Pan Stanford Publishing Pte. Ltd.
ISBN 978-981-4411-25-7 (Hardcover), 978-981-4411-26-4 (eBook)
www.panstanford.com

12.2   Experiment

In order to compare the measurement by the Kerr effect–based system, the classical Sixtus–Tonks setup was used. Two pickup coils are separated by 6 cm. More details of the system can be found in [1]. The measurements were performed on the amorphous glass-coated microwire of the composition $Co_{68}Mn_7Si_{15}B_{10}$. The microwire was 10 cm long, with the diameter of metallic nucleus of 8 μm and total diameter of 20 μm. This microwire has very low but positive magnetostriction. It is just high enough to show magnetic bistability. It shows magnetic bistability in the temperature range from low temperature up to 350 K. Above this temperature, the magnetic bistability disappears. This wire was chosen due to its low anisotropy (which results from its low magnetostriction) that guarantees low domain wall damping and therefore high domain wall velocity.

In order to compare the measurement by the Kerr effect–based system, the classical Sixtus–Tonks setup was used. Two pickup coils are separated by 6 cm. More details of the system can be found in [1]. The measurements were performed on the amorphous glass-coated microwire of the composition $Co_{68}Mn_7Si_{15}B_{10}$. The microwire was 10 cm long, with the diameter of metallic nucleus of 8 μm and total diameter of 20 μm. This microwire has very low but positive magnetostriction. It is just high enough to show magnetic bistability. It shows magnetic bistability in the temperature range from low temperature up to 350 K.

12.3   Comparison of Induction and MOKE Methods

Figure 12.1 shows the signal obtained both from the classical and Kerr effect-based Sixtus–Tonks experiments. Average function is used to obtain almost "pure" signal where only voltage drops arising from the domain wall propagation through the reflection point are recognized. The time of the domain wall propagation between the two reflection points can be clearly recognized. The disadvantage of the induction method is that the pickup coil width must be taken into account (3 mm in our case), whereas the width of the laser beam is much smaller (0.5 mm). Moreover, the width of the maxima induced in the pickup coil does not decrease with the increase of the velocity as a result of its relaxation time.

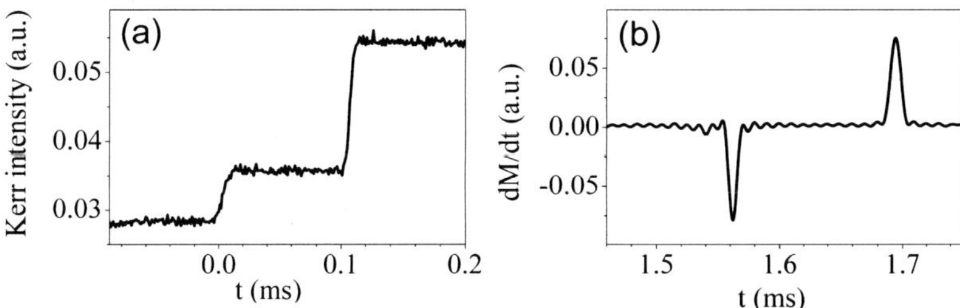

Figure 12.1 An example of the Kerr effect signal (a) and the signal obtained by the classical induction (b) Sixtus–Tonks experiment.

On the other hand, the change of the signal in the case of Kerr effect–based experiment is sharper. The smooth change at the first reflection point is a result of the averaging and the switching field fluctuation [2]. However, further improvement of the method is in progress, which allows us to measure even higher domain wall velocity, when the time, which the domain wall needs to pass between two positions, is shorter than the relaxation time of the pickup coils.

Figure 12.2 shows field dependence of the domain wall velocity measured at room temperature by the classical and Kerr effect–based Sixtus–Tonks experiments. According to ref. [3], the field dependence of the domain wall velocity is expressed as follows:

$$V = S(H - H_0), \tag{12.1}$$

where S is the domain wall mobility and H_0 is the so-called critical propagation field, below which the domain wall propagation is not possible. Such linear dependence is confirmed at high field. However, the critical field H_0 is negative. Similar negative critical field has already been found in amorphous microwires [1, 3] and could be explained by the negative nucleation field of the closure domain wall. Anyway, the domain wall velocity is quite high and reaches up to 900 m/s.

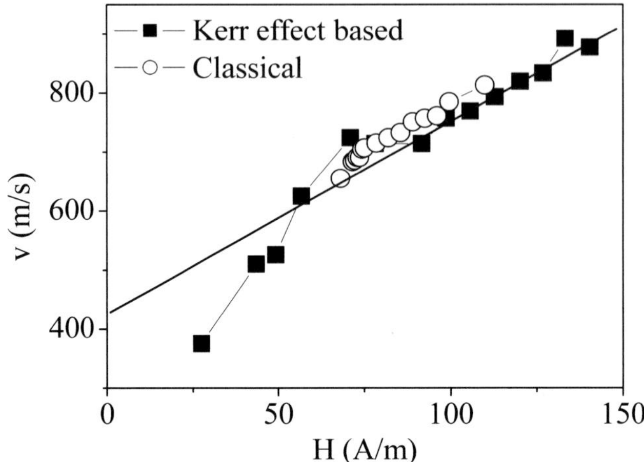

Figure 12.2 Domain wall velocity *v* as a function of the magnetic field amplitude *H* for amorphous glass-coated CoMnSiB microwire measured by the classical and by the Kerr effect–based Sixtus–Tonks experiments. Full line is a fit according to Eq. (12.1).

At low field (below 75 A/m), the change of the domain wall mobility *S* is found. In this range, the domain wall dynamics change from the viscous motion to adiabatic one. However, at low fields, the local fluctuation of the short-range interaction of the domain wall with the defects plays its role and the domain wall propagates in intermittent jumps with the velocity that fluctuates according to the local pinning field. The average velocity of the domain wall finally scales as

$$V = S'(H - H_0')^\beta,$$
(12.2)

where S' is the so-called domain wall mobility parameter and H_0' is the critical field.

12.4 Domain Walls Dynamics in Co-Rich Microwires

During these experiments, we paid attention to the influence of the external DC axial magnetic field on the surface circular DW motion because the influence of this magnetic field on the impedance plays the essential role in the GMI effect. Accordingly, an experiment with increasing value of the DC axial magnetic field has been realized. The pulsed circular magnetic field induced the DW motion and was fixed on the value of 0.2 Oe (Fig. 12.3). First, the series of the time dependencies of the MOKE jumps related to the surface DW motion along the wire has been obtained and then the dependence of the circular DW velocity on the external DC axial field has been plotted (Fig. 12.6). The general growth of the velocity with the DC axial field is observed. The value of the DW velocity of about 2 km s^{-1} has been achieved in this experiment.

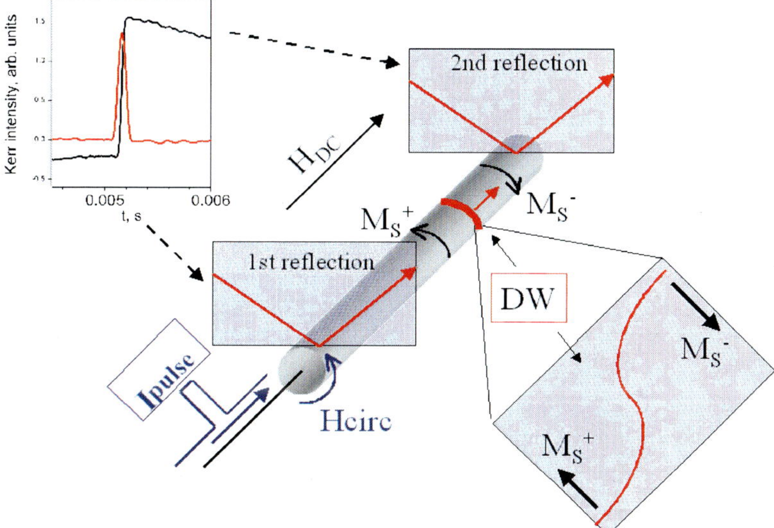

Figure 12.3 Schematic picture of MOKE modified Sixtus–Tonks method. Inset shows MOKE signal jump related to domain wall motion along the laser spot (black line) and its derivative (red line).

Three specific parts of the obtained dependence were marked out. There are two parts with clearly defined increase of the velocity (marked as "I" and "III" in Fig. 12.4)—at the beginning and at the end of the curve, respectively. Also, the local decrease of the velocity exists in the middle part of the field dependence of the DW velocity (marked as "II").

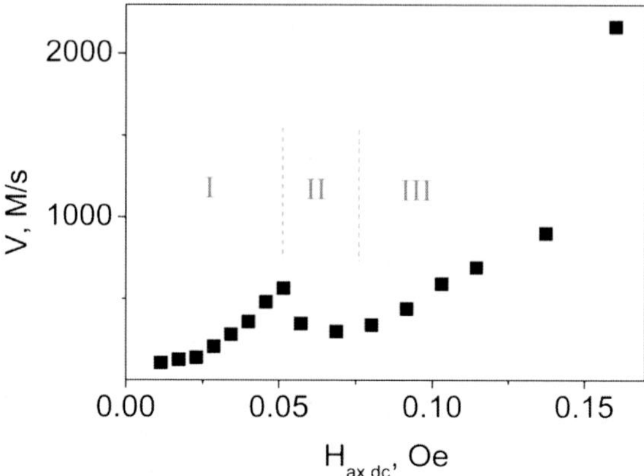

Figure 12.4 Dependence of surface circular domain wall velocity on DC axial field. The constant value of driving pulsed circular magnetic field is 0.2 Oe.

The obtained dependence has been analyzed taking into account the influence of the DC axial magnetic field on the shape of the MOKE jump related to the passing of the DW across the laser spot (Fig. 12.5). The main idea of the analysis is that the time dependence of the MOKE signal (MOKE jump) contains the information about the magnetic structure of the DW moving along the laser spot. The similar idea has been used for the analysis of the domain wall motion in Fe-rich magnetic microwires [4]. In other words, the direct correspondence between the time profile of MOKE signal and the space profile of DW exists in the performed experiments.

There are three stage of the MOKE jump transformation depending on the value of the axial field. In the first stage (marked as "I"), the time duration of the jump shortens with axial field increase (Fig. 12.5a). In the second stage (marked as "II"), the specific transformation of the shape of the jump was observed (Fig. 12.5b) the jump derivative has two peaks instead of one observed in stage "I." We consider that this effect is related to the transformation of the magnetic profile of the DW. It will be discussed below. Finally (stage "III"), the absolute value of the MOKE jump decreases (Fig. 12.5b) and the MOKE signal totally disappears when the DC field reaches some special value. This disappearance is reasonable because the DC axial field of the relatively high value directs the magnetization in two surface domains along the axial direction of the microwire and in that way simultaneously eliminates the circular projection of the surface magnetization. This rotation of the magnetization causes the disappearance of surface circular domain wall.

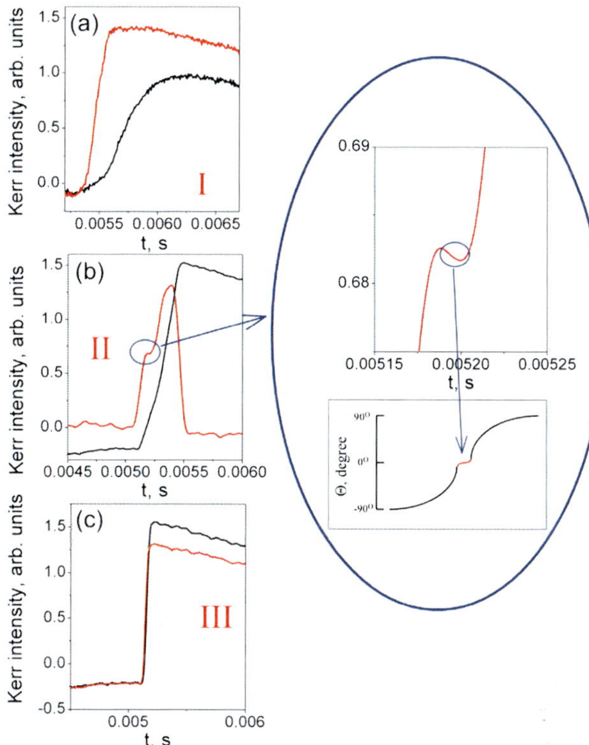

Figure 12.5 DC axial magnetic field induced transformation of the shape of the MOKE jump. Time dependencies of MOKE signal corresponding to the passing of the DW across the laser spot. Stage I: (a) H_{AX} = 0.02 Oe (black line), H_{AX} = 0.03 Oe (green line); stage II: (b) H_{AX} = 0.07 Oe, black line—MOKE jump, red line—derivative of MOKE jump; stage III: (c) H_{AX} = 0.12 Oe (black line), H_{AX} = 0.16 Oe (blue line). Inset shows the correspondence between the MOKE signal and the magnetic profile in DW for the stage II.

We consider that the strong correlation exists between the DC field dependence of the DW velocity (Fig. 12.4) and the axial field–induced transformation of the MOKE jump (Fig. 12.5). The schematic pictures of the tree stages of the DW transformation are presented in Fig. 12.6 and marked there as I, II, and III.

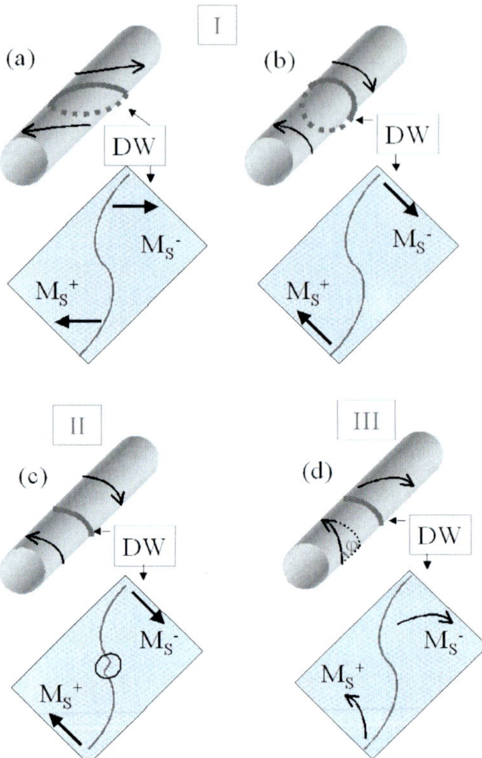

Figure 12.6 Schematic pictures of magnetic profile of domain wall for tree stages of DW transformation. (a) Inclined DW, (b) perpendicular DW, (c) DW which contains extended part with magnetization parallel to axial magnetic field, (d) DW between two domains with magnetization partially rotated towards axial direction.

The axial field–induced decrease of the time duration of the MOKE jump observed in a relatively small value of the DC magnetic field is related to the DW transformation from the inclined one to the DW perpendicular to the microwire axis. This effect has the same nature as the axial field–induced transformation of surface helical magnetic structure presented earlier in our paper [5]. The inclined DW divides two helically magnetized surface domains. The inclined domain walls have been found in Co-rich glass-covered microwires using the MOKE polarizing microscope [6].

The schematic pictures of the inclined and the perpendicular domain walls are presented in Figs. 12.6a,b. In mentioned range of axial fields, the DC axial field induces the rotation of the DW to perpendicular direction and as a consequence, the DW length becomes smaller. In supposition of the thermo-activated mechanism of DW motion [7], the domain wall with smaller length has higher mobility because it interacts with smaller number of defects.

In the second stage of the DW transformation, the velocity partially decreases when the DC axial field continues to increase. The MOKE jump and its derivative presented in Fig. 12.1b correspond to this part of the DC field dependence of the velocity. The derivative has specific shape of clearly divided two peaks in this stage "II." These two peaks are observed for all values of the DC field that corresponded to stage "II." This means that the DC axial field induces the specific transformation of the DW in this stage (Fig. 12.6c). Two peaks correspond to two successive rotations of the magnetization from the circular direction "+" to the axial direction and then from the axial direction to the circular direction "−" inside the DW. A small part between two peaks (see inset in Fig. 12.5) corresponds to the part inside DW with magnetization directed along the axial direction. Thus, the kink observed in the derivative of the MOKE signal (Fig. 12.5b) is the clear confirmation of the existence of extended part with the magnetization parallel to the DC axial field inside the surface DW.

The DW has an ordinary shape when the DC axial field is small (stage "I"). It corresponds to the fluent, 180° rotation of the magnetization inside the DW (Fig. 12.6b). When the DC axial field is relatively high (stage "II"), the domain wall contains extended part with the magnetization parallel to the axial magnetic field [7] (marked by black circle in Fig. 12.6c). As it was mentioned earlier, the surface DW motion is induced by the circular magnetic field. Therefore, being perpendicular to the driving circular field, this part of the DW plays the key role in the deceleration of the DW motion when the driving pulse circular field is maintained at the constant value as it was performed in the present experiments.

The third stage of the DW transformation is illustrated schematically in Fig. 12.6d. This transformation is related to the final acceleration of the DW motion observed in **Fig. 12.4** beginning from the value of the axial magnetic field of about 0.8 Oe (stage III). In this stage, the axial magnetic field noticeably rotates the magnetization toward the axial direction in two surface magnetic states. The angle between the surface magnetization and the axial direction is marked as "φ" in Fig. 12.6d. This angle decreases when the DC axial field is high enough. The angle of the rotation of the magnetization inside the DW $\theta = 2\varphi$. It is known that the DW velocity depends directly on the value of the angle of the rotation of the magnetization inside the domain wall. The decrease of this angle causes the increase of the DW velocity. The magnetization rotating with the DC axial field toward the axial direction reaches the axial direction at some value of the DC field. The surface circular DW disappears at this moment.

It is necessary to note that the shape of the derivative of the MOKE signal of two peaks is not observed in stage III. The derivative has only one peak. This means that the extended part of the DW with the magnetization perpendicular to the axial magnetic field disappears when the angle θ considerably decreases. In other words, the process of the axial field–induced reduction of the angle θ suppresses the process of the extension of the special part with kink inside the DW.

Therefore, a strong influence of the DC axial magnetic field on the surface circular domain wall motion was found. Also, the axial field–induced transformation of the magnetic profile of the moving DW was discovered. The correlation between these two processes was demonstrated experimentally. The DC axial field–induced transformation of DW has three stages: (I) the rotation of the whole DW from the inclined to the circular direction, (II) the appearance of extended part inside the DW with magnetization directed along the DC axial field, and (III) the decrease of the angle of the rotation of the magnetization inside the DW. Processes I and II cause the acceleration of the DW motion, while process II decelerates it. As a result of the competition of these three processes, depending on which effect predominates the DC axial field acts as accelerator or decelerator of the surface circular DW motion. Therefore, we have found that using the DC axial magnetic field as the unique external parameter, we can control the surface circular DW motion, its shape, and, as a result, the moment of its annihilation.

References

1. Varga R, Zhukov A, Blanco JM, Ipatov M, Zhukova V, Gonzalez J, and Vojtaník P (2006), *Phys. Rev. B*, **74**, 212405.

2. Varga R, Garcia KL, Zhukov A, Vazquez M, and Vojtanik P (2004), *Phys. Rev. B,* **70**, 024402.

3. Varga R, Zhukov A, Usov N, Blanco JM, Gonzalez J, Zhukova V, and Vojtanik P (2007), *J. Magn. Magn. Mater.,* **316**, 337.

4. Gudoshnikov SA, Grebenshchikov YuB, Ljubimov BYa, Palvanov PS, Usov NA, Ipatov M, Zhukov A, and Gonzalez J (2009), *Phys. Status Solidi A,* **206**, 613.

5. Chizhik A, Zablotskii V, Stupakiewicz A, Gómez-Polo C, Maziewski A, Zhukov A, Gonzalez J, and Blanco JM (2010), *Phys. Rev. B,* **82**, 212401.

6. Gonzalez J, Chizhik A, Zhukov A, and Blanco JM (2011), *Phys. Status Solidi A,* **208**, 502.

7. Varga R, García KL, Zhukov A, and Vazquez M (2004), *Phys. B: Cond. Mat.,* **343**,

Chapter 13

Nucleation and Transformation of Circular Magnetic Domain Structure: Control of Domain Nucleation

13.1 Introduction

This chapter is devoted to study of surface magnetic domains nucleation and domain wall motion related to it in Co-rich microwires. The measurements were performed by the MOKE polarizing microscope and the MOKE-modified Sixtus–Tonks method.

13.2 Circular Field–Induced Nucleation and Transformation of Circular Magnetic Domains

The process of magnetization reversal in the surface area of the microwires has been studied in a microwire of nominal composition $Co_{67}Fe_{3.85}Ni_{1.45}B_{11.5}Si_{14.5}Mo_{1.7}$ (metallic nucleus radius 11.2 μm, glass coating thickness 3 μm) in the presence of circular magnetic field (Fig. 13.1). The circular magnetic domains could be observed because of the out-of-plane components of the surface magnetization that transforms to black-white contrast when the polarized light reflects from the cylindrical-shape surface of the microwire (Fig. 13.2).

Magnetic Microwires: A Magneto-Optical Study

Alexander Chizhik and Julian Gonzalez

Copyright © 2014 Pan Stanford Publishing Pte. Ltd.

ISBN 978-981-4411-25-7 (Hardcover), 978-981-4411-26-4 (eBook)

www.panstanford.com

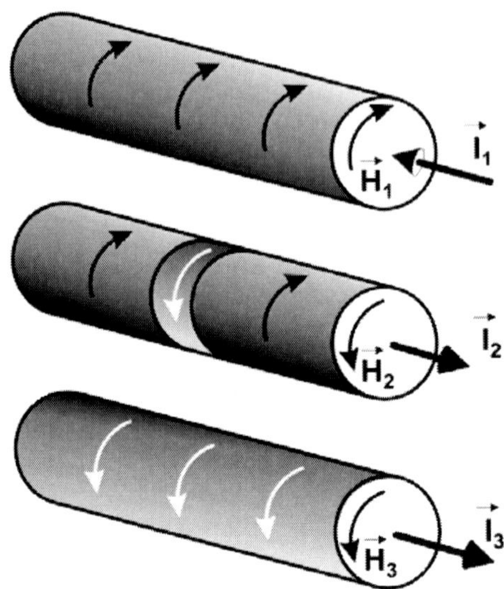

Figure 13.1 Schematic picture of the circular domain nucleated by circular magnetic field in microwire.

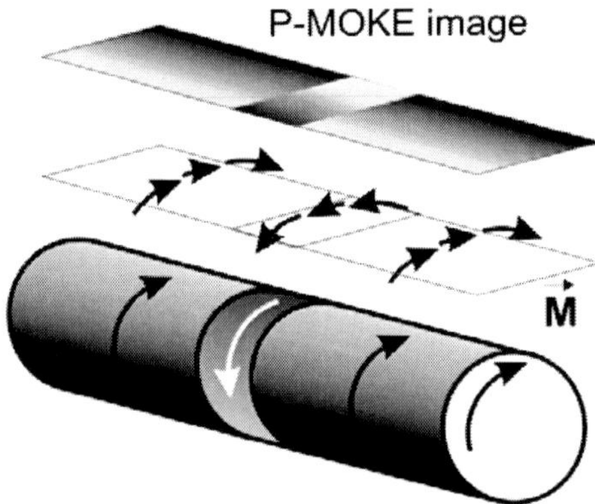

Figure 13.2 Configuration of the circular domain structure image observation using polar magneto-optical Kerr geometry.

The results of the polar-MOKE microscopy experiments are presented in Fig. 13.3. The magnetization reversal begins from the mono-domain state (Fig. 13.3a). Notably, one of the interesting results obtained is that the magnetization reversal could start from practically any part of the microwire.

In the first stage of the magnetization reversal, the nucleation and fast domain wall motion causes the formation of the relatively small circular domains (Fig. 13.3b). The following increase of the circular magnetic field leads to the increase of the number of the domains with practically equal width (Fig. 13.3c). In the next stage, the strong rearrangement of the domain structure takes place: The number of the domains increases sharply (Fig. 13.3d). The magnetization reversal finishes as the reversed mono-domain state.

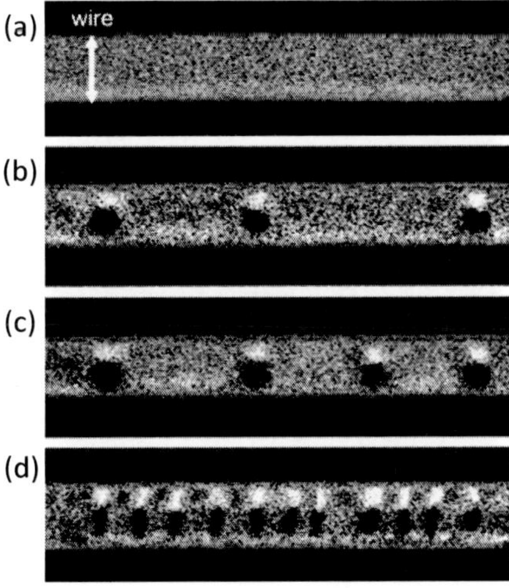

Figure 13.3 The images of magnetic domain structure registered in different values of circular magnetic field: (a) 0, (b) 0.3 Oe, (c) 0.36 Oe, (d) 0.45 Oe. Image size is 170 μm × 50 μm.

The magnetization reversal process could be divided into two parts: relatively short part associated with the domain wall propagation and the following part determined basically by the domain nucleation process. The obtained results could be related to the thermo-activated mechanism of the domain nucleation and domain wall motion [1]. The competition of the energetic barriers related to the domain wall motion and domain nucleation determines mainly the observed magnetization reversal. For relatively high magnetic field, the nucleation-dominated reversal takes place. In this stage, the local distribution of the nucleation centers density is the key parameter of the magnetization reversal [2, 3]. This distribution depends strongly on the magnetic field force. As it is possible to see in our experiment, domain wall motion is limited by the high pinning of the domains walls and the number of the nucleation centers growth dramatically with the magnetic field increase. The growth of the number of the reversed domain could be divided into two stages. In the first stage, the simple increase of the circular domains number is observed (Figs. 13.3b,c). In the second stage (Fig. 13.3d), the number, positions, and wideness of domains change as a specific form of the magnetic structure reorganization [4]. Domain wall motion is limited by the high pinning of the domains walls.

Circular magnetic domains nucleation process in microwires could be qualitatively described using thermally activated mechanism for number of domain nucleation N [1]:

$$N(H) \sim \exp\frac{\Delta E_n}{kT}(H - H_n), \tag{13.1}$$

where ΔE_n is the nucleation energy barrier, H is applied circular magnetic field, H_n is the starting field of domain nucleation. H_n was determined from the linear dependence in Fig. 13.4 (d is the sum of the widths of the reversed circular domains in the analyzed image, d_s is the width of the microwire in the analyzed image).

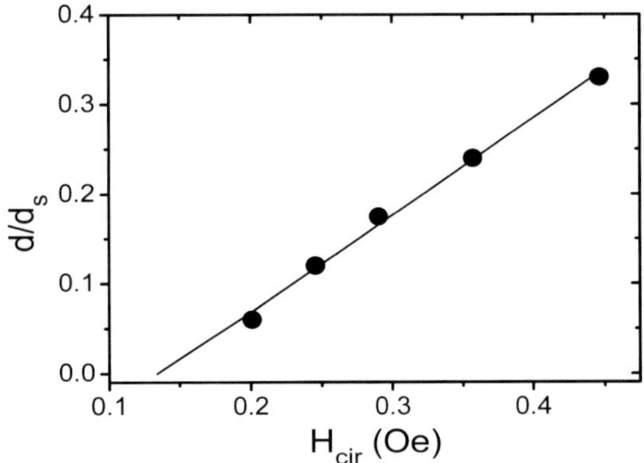

Figure 13.4 Magnetic field dependence of the normalized sum of the widths of circular domains.

The functional dependence of Eq. (13.1) shows the exponential growth of number of nucleation centers with respect to H in a good agreement with experimental data given by points (see Fig. 13.5).

The formation of circular multi-domain structure in the Co-rich microwires is related to the strong correlation between the sign and value of the magnetostriction constant and the type of domain structure [5, 6]. The MOKE magnetometry experiments in the presence of the external stresses have shown that the circular multi-domain structure exists in determined range of stresses and that the application of the external tensile stress could transform the multi-domain structure to mono-domain structure and initiate appearance of the circular bistability effect [7]. When the external stresses are small enough, the stability of multi-domain circular structure in the nearly-zero magnetostrictive composition could be reasonable because of the competition between the magnetostatic and the magnetoelastic energy at the condition when the magnetoelastic energy is low enough. The circular magnetic bistability is related to the magnetoelastic anisotropy in the circular direction, as the classical longitudinal magnetic bistability [8]. The circular magnetoelastic anisotropy results in the appearance of circular domain structure in the outer shell.

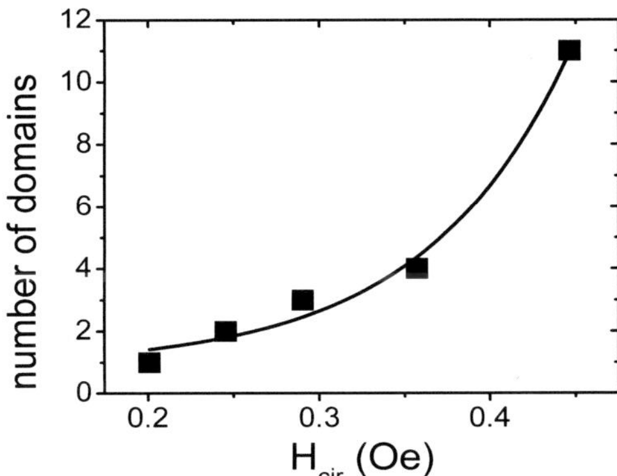

Figure 13.5 The dynamics of nucleation domain number versus circular magnetic field.

In the absence of the external tensile stress, the key influence on the character of the surface domain structure in the glass-covered microwires exerts the internal stresses originated by the glass covering [9]. The internal stresses are produced due to the contraction of the metal and glass having different thermal expansion coefficients [10].

In the frame of the core–shell model, which considers that a Co-based amorphous wire has an inner core with longitudinal easy axis and an outer shell with circular or helical anisotropy [11], we have found earlier the strong correlation between the direction of the circular magnetization in the outer shell and the direction of axial magnetization in the inner core. Therefore, taking into account this correlation, we could suppose that the multi-domain structure also forms in the axially magnetized inner core during the studied magnetization reversal.

13.3 Control of Domain Nucleation

To study the domain wall motion, two reflections of the broken laser beam from the microwire surface were used instead of the pickup coils used in Sixtus–Tonks experiments. Two points of reflection are separated by 6 cm.

In the first stage of our experiments, to study the influence of the nucleation field on the switching field, we used the experimental scheme of one reflection. The switching field associated with the domain wall motion along the wire was registered as a jump of the Kerr signal. The correlation between the switching field (H_{SW}) and the magnetic field in the nucleation coil (H_n) was carried out. Figure 13.6 presents the strong dependence of the value of H_{SW} on the amplitude of H_n: significant decrease of the switching field takes place as a result of the growth of the field in the nucleation coil.

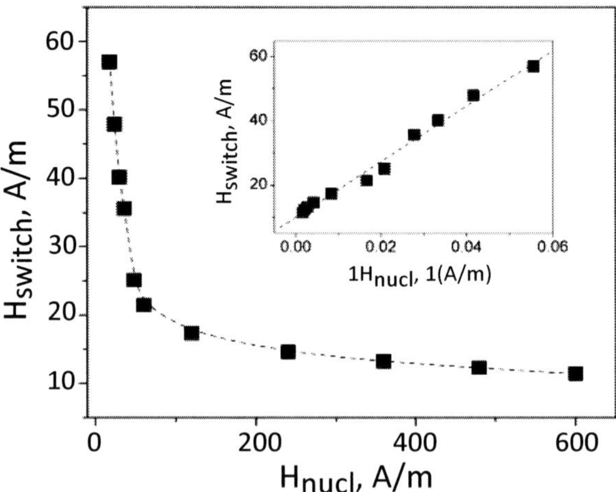

Figure 13.6 Dependence of switching field H_{SW} on nucleation field H_n. In the inset the experimental points are re-plotted as a dependence on $1/(H_n)$.

Analyzing the obtained dependence of the switching field we apply the idea that the growth of the magnetic field in the nucleation coil could be considered as the analogy of a temperature increase. In other words, in the assumption of the thermo-activated mechanism of the domain nucleation and domain wall motion, an increase of the amplitude of the magnetic field initiating the domain nucleation and the increase of the temperature cause, in similar way, the increase of the probability of the overcoming of the energetic barrier related to the domain nucleation.

Following [12–14], in which the relation between the switching field and the temperature was presented as $H_{SW} \sim F/(M_S T)$, (where F is a relaxation function and M_S is saturation magnetization) we consider that the relation of the H_{SW} and the H_n could be mainly determined by the expression $H_{SW} \sim 1/H_n$. The experimental results have been re-plotted as dependence H_{SW} $(1/H_n)$ (see inset to Fig. 13.6). The experimental points lie very good on the linear dependence that demonstrates the existence of $H_{SW} \sim 1/H_n$ correlation.

The nucleating field increases the probability of the domain nucleation and in turn, decreases the switching field. We have experimentally shown this. Since the property of the energetic barrier is not affected by the nucleating field, we can track separately the magnetic domain nucleation process in the conditions of the unchanged energetic barrier.

To study the possibility of the control of the domain nucleation, we varied the phase of the magnetic field in the nucleation coil jointly with the application of the DC electric current flowing along the microwire.

First, we studied the influence of the change of the phase in the nucleation coil. It was found that the phase shift Δ causes the time shift of the switching field or totally suppresses the domain nucleation (Fig. 13.7). The monotonic decrease of the time delay of the domain nucleation takes place in the interval of 0–150°. The magnetization reversal and domain wall motion were not observed in the phase interval of 150–180°. These results look reasonable taking into account that the phase shift causes the shift of the moment when the dependence of the nucleating field crosses the "0" line (see inset to Fig. 13.7). The domain nucleation could appear only after the moment when the positive growth of the nucleation field starts. When the phase shift is close to 180° the nucleation field it in (or almost in) anti-phase to the external magnetic field. In these conditions the nucleation process is suppressed.

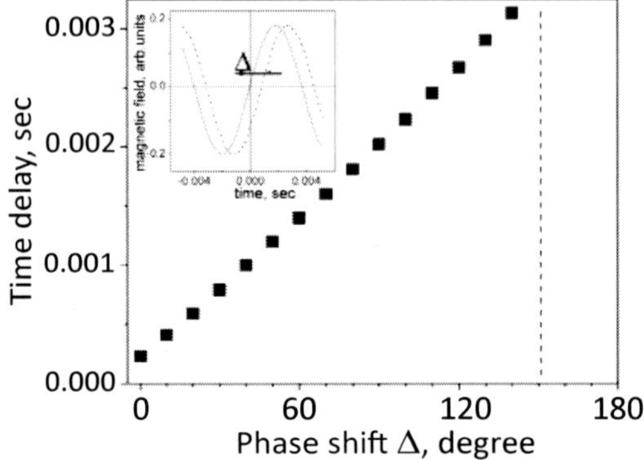

Figure 13.7 Dependence of the time delay of switching field on phase shift Δ in nucleation coil. The inset shows schematically the distribution of the magnetic fields in the Helmholtz coils and the nucleation coil for the phase shift Δ.

Significant changes of the magnetization reversal have been observed when we additionally applied an electric current along the microwire. Earlier, the study of the influence of the electric current and the circular magnetic field produced by the current on the magnetization reversal has been performed and demonstrated strong correlation between the circular magnetic field and the switching field [15–17].

At this stage of the experiments, we used the two-spot configuration mentioned above (see the scheme in Fig. 13.8): When the domain wall related to the magnetization reversal moved along the wire, two jumps of the Kerr effect signal took place. The magneto-optic scheme has been adjusted in such a way that when the domain wall has passed through the laser spots, the jump "up"-down" (for the spot "1") or jump "down"-"up" (for the spot "2") has been observed (two analyzers for two laser beams have been inclined to different directions from the crossing position).

Figure 13.8a shows the time dependence of the Kerr signal in spots 1 and 2 when the phase shift was 0 and the electric current was not applied to the microwire (circular magnetic field is 0). The domain wall motion begins from the end of the microwire where the nucleation coil was placed: First the domain wall goes along spot 1 and then along spot 2. The joint application of the phase shift and electric current cardinally changes the magnetic configuration of the experiment (Fig. 13.8b). The jump in spot 2 is observed first and then the domain wall moves along spot 1. This time distribution of the Kerr effect jumps demonstrates that the domain wall begins the motion from the opposite end of the microwire. This effect is observed in the wide enough interval of phase shift (40–130°).

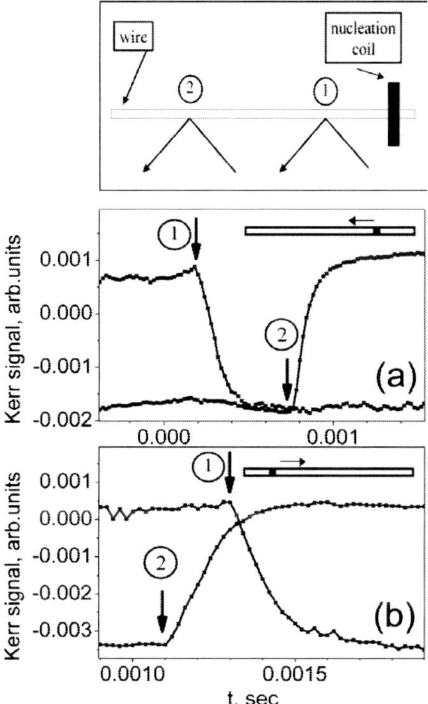

Figure 13.8 Time dependencies of Kerr signal in spot 1 and spot 2 for two different experimental configurations: (a) domain wall starts from the end of the microwire where the nucleation coil was placed; (b) domain wall starts from the opposite end of the microwire. The positions of the nucleation coil and two laser spots are shown in the schematic picture.

Generally, the value of the domain wall velocity could be measured using the two-spot scheme. However, the present version of the experimental configuration permits us to perform only the comparative analysis of the velocity. If we consider the points marked by the arrows in Figs. 13.8a,b as the moments when the domain wall starts to moves along the light spot, it is possible to see that the time delay between two jumps of Kerr signal is shorter for the second case (Fig. 13.8b), and, respectively, the velocity of the domain wall is higher. The observed increase of the domain wall velocity could be related to the DC current influence. The circular magnetic field related to the DC current causes the change of the angle of the rotation of the magnetization in the domain wall. The domain wall with smaller rotation angle moves faster. On the other hand, special attention has to be paid to the time duration of the transition region of the Kerr signal (from "up" to "down" or from "down" to "up")—from the moment when the domain wall reaches light spot region and up to the moment when the domain wall leaves it. This transition region is longer for the case of Fig. 13.8b than for the case of Fig. 13.8a, which could be also related to the change of the value of the domain wall velocity. The reason for this effect of circular magnetic field (DC electric current)—induced change of the transition region of the Kerr signal is not currently clear. The precise measurements of the domain wall velocity need some changes in the experimental setup (make the light spot smaller, application of the pulse magnetic field with rectangular shape of the pulse, etc.), which we are going to perform.

Two correlated effects could be proposed as a possible reason for the observed phenomenon of the controlled change of the direction of the domain wall propagation. As we can see above, the phase shift could be considered as a decelerator of the domain nucleation that finds the realization in the time delay of the start moment of the domain wall motion. In the same time, the circular magnetic field serves as the accelerator of the axial magnetic domain nucleation. As a result, the change of the direction of the domain wall movement comes out from the superposition of two effects with opposite actions.

The circular magnetic field produces the inclination of the magnetization from the axial direction that induces and accelerates the opposite domain nucleation in the whole microwire. Therefore, two opposite ends of the microwire are under the similar influence of the circular magnetic field. The magnetic field induced by the nucleation coil in some phase shift suppresses and delays the nucleation process in one end of the microwire and does not affect another end of the microwire. In this situation, the nucleation and the following domain wall motion were initiated at the end of wire where the nucleation coil is not placed (in our scheme—the left end). The key detail of this process is that the phase shift produces enough long time delay of the nucleation in the right end of the microwire keeping the time interval enough long for the domain nucleation in left end.

Manipulating the amplitude and the phase of the magnetic field in the nucleation coil jointly with the DC electric current applying to the microwire we can accelerate, decelerate and suppress the domain nucleation process. Also, the place of the domain nucleation could be selected artificially selecting the specific configuration of the above-mentioned parameters. The influence of the amplitude of the nucleation field on the switching field could be basically considered as an analogy of the temperature influence in the frame of the model of the thermo-activated nucleation and domain wall motion. On the basis of this, the H_{SW} (H_n) dependence could used to predict the temperature changes of the magnetization reversal and temperature dependencies of switching field in magnetic microwires.

References

1. Vogel J, Moritz J, and Fruchart O (2006), *C. R. Phys.*, **7**, 977.

2. Pommier J, Meyer P, Penissard G, Ferre J, Bruno P, and Renard D (1990), *Phys. Rev. Lett.*, **65**, 2054.

3. Davies JE, Hellwig O, Fullerton EE, G. Denbeaux G, Kortright JB, and Liu K (2004), *Phys. Rev. B*, **70**, 224434.

4. Hubert A and Shaefer R (1998), *Magnetic Domains* (Berlin: Springer).

5. Humphrey FB, Mohri K, Yamasaki J, Kawamura H, Malmhall R, and Ogasawara I (1987), in: A. Hernando, *et al.*, (Eds.), *Magnetic Properties of Amorphous Metals* (Elsevier, Amsterdam).

6. Chizhik A, Zhukov A, Blanco JM, and Gonzalez J (2003), *J. Magn. Magn. Mater.*, **254–255C**, 188.

7. Chizhik A, Gonzalez J, Zhukov A, and Blanco JM (2003), *J. Phys. D: Appl. Phys.*, **36**, 419.

8. Mohri K and Takeuchi S (1982), *J. Appl. Phys.*, **53**, 8386.

9. Chizhik A, Blanco JM, Zhukov A, Gonzalez J, Garcia C, Gawronski P, and Kulakowski K (2006), *IEEE Tran. Magn.*, **42**, 3889.

10. Velasquez J, Vazquez M, and Zhukov A (1996), *J. Mater. Res.*, **11**, 2499.

11. Vazquez M and Hernando A (1996), *J. Phys. D.,* **29**, 939.

12. Ponomarev BK and Zhukov AP, (1985), *Sov. Phys. Solid State* **27**, 272.

13. Zhukova V, Zhukov A, Blanco JM, Gonzalez J, and Ponomarev BK (2002), *J. Magn. Magn. Mater.,* **249**, 131.

14. Varga R, Garcia KL, Vazquez M, Zhukov A, and Vojtanik P (2004), *Phys. Rev. B,* **70**, 024402.

15. Chizhik A, Gonzalez J, Zhukov A, and Blanco JM (2002), *J. Appl. Phys.,* **91**, 537.

16. Chizhik A, Garcia C, Gonzalez J, and Blanco JM (2004), *J. Magn. Magn. Mater.,* **279**, 359.

17. Gonzalez J, Chizhik A, Zhukov A, and Blanco JM (2003), *J. Magn. Magn. Mater.,* **258–259**, 177.

Chapter 14

Magnetization Reversal in Co-Rich Microwires with Different Values of Magnetostriction

14.1 Introduction

This chapter is devoted to the Kerr effect study of the magnetization reversal at the surface of the glass-covered Co-rich microwires with different value and sign of the saturation magnetostriction constant. Magnetic and mechanical properties of these wires were studied in Ref. [1], in which a dramatic dependence of magneto-strictive properties on the content of Mn was demonstrated.

14.2 Experimental Results and Discussions

The Kerr effect experiments were performed on three wires with different Mn content (nominal composition $(Co_{1-x}Mn_x)_{75}Si_{10}B_{15}$ ($x = 0.07, 0.09, 0.11$)).

There were four schemes of the experiments, depending on the combination of magnetic field combination and the type of Kerr effect:

Magnetic Microwires: A Magneto-Optical Study
Alexander Chizhik and Julian Gonzalez
Copyright © 2014 Pan Stanford Publishing Pte. Ltd.
ISBN 978-981-4411-25-7 (Hardcover), 978-981-4411-26-4 (eBook)
www.panstanford.com

(1) transverse Kerr effect at sweeping of circular field (± axial bias field);

(2) longitudinal Kerr effect at sweeping of circular field (± axial bias field);

(3) transverse Kerr effect at sweeping of axial field (± circular bias field);

(4) longitudinal Kerr effect at sweeping of axial field (± circular bias field).

The Kerr effect loops obtained for the microwire with $x = 0.07$ are presented in Fig. 14.1. The rectangular shape of the transverse curve is obtained in AC circular field (Fig. 14.1a). It is worth to note that the minor loop was not observed below some AC circular field, and the rectangular hysteresis loop abruptly appeared when the circular magnetic field reached some critical value. These properties are attributed to the single and large Barkhausen jump [2].

The longitudinal hysteresis loop in AC circular field exhibits two peaks (Fig. 14.1b). Figures 14.1c–g present the transverse Kerr curves, which were obtained in the AC axial field, under the DC circular field. The values of DC electric current flowing through the wire and producing DC circular magnetic field are marked in the figure. The comparison of transverse and longitudinal loops permits us to clarify the mechanism of magnetization reversal of the wire. We assume an existence of circular bamboo domains in the studied wire. The rectangular transverse Kerr loop (Fig. 14.1a) and peaks in longitudinal Kerr loop (Fig. 14.1b) are related to quick change of magnetization between two circular directions. The transformation of the transverse Kerr curves obtained in the AC axial field (Figs. 14.1c–g) with the circular magnetic field as a parameter, confirms the existence of circular domain structure in outer shell of the wire. When the DC circular magnetic field is applied, the volume of circular domains of one type increases and of another type decreases, which is displayed in the curve transformation. Figures 14.1c,g show the rotation of magnetization only in the one domain ("+" or "−" depending of the sign of the DC circular field). Relatively quick change of the sign of the MO signal (marked by the arrows in Figs. 14.1d,f could be associated only with the increasing of the volume of circular domains of one type and the decreasing of the volume of circular domains of another type. Before and after this quick change of MO signal, the relatively slow rotation of magnetization was observed [3–5].

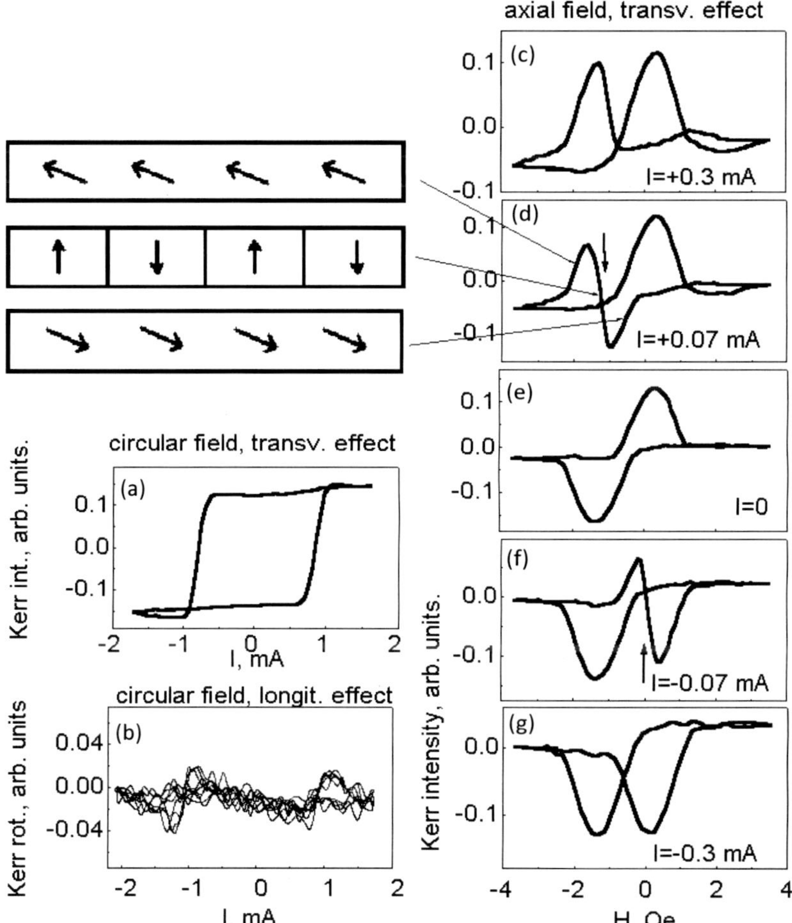

Figure 14.1 Kerr effect loops of Co-rich wire with $x = 0.07$. Inserted picture presents schematically the proposed domain structure in the outer shell of the microwire.

For the microwire with content of Mn $x = 0.11$, the rectangular shape of the Kerr loop was found under the AC axial field (Fig. 14.2a). Sharp change of Kerr effect in longitudinal curve is attributed to a quick change of the axial magnetization. In contrast with in the microwire with $x = 0.07$, the transverse curve under AC circular field of the microwire with $x = 0.11$ (Fig. 14.2b) demonstrates the smooth shape related to the fluent rotation of the magnetization. The shapes of the observed curves were analyzed using the model, which takes into account existence of a maze domain structure with closure domains in the outer shell of the wire. It is known that the magnetization in the closure domains is aligned parallel to the axis of the wire. The transformation of transverse loops obtained in the AC axial field, in presence of the DC circular field (Figs. 14.2c–g) could clarify details of the domain structure. The absence of the MO signal when the DC circular field is zero can be explained by the absence of circular component during remagnetization. That is, the magnetization reversal occurs only by the increase in the volume of longitudinal domains of one type and the decrease in the volume of longitudinal domains of another type. The application of the DC circular field produces a deviation of the magnetization in the domains from the axial direction and the rotation of magnetization takes place additionally to the domain walls motion. The value of the transverse projection of the magnetization increases with the DC circular field. This is reflected with the increase of value of the MO Kerr signal (comparing Figs. 14.2c,d).

Earlier, our investigation of glass-coated $(Co_{1-x}Mn_x)_{75}Si_{10}B_{15}$ amorphous microwires performed by the method of variation of the initial permeability with the applied tensile stress [1] allowed us to conclude that the magnetostriction constant λ_S is positive for wires with $x \geq 0.1$ and negative for wires with $x < 0.1$. Taking into account that the circular magnetization alignment in the outer shell is attributed to the negative sign of the magnetostriction and axial magnetization alignment in the outer shell—to positive magnetostriction [2], the modification of the domain structure of glass-coated $(Co_{1-x}Mn_x)_{75}Si_{10}B_{15}$ amorphous microwires could be ascribed to the change of the value and sign of the magnetostriction constant with x. Thus, our magneto-optical investigation for the microwires with $x = 0.07$ and for $x = 0.11$ are in agreement with independent evaluation of the magnetostriction constant value. It is worth to note that the rectangular transverse hysteresis loop discovered in microwire with $x = 0.07$ could be attributed to single and large Barkhausen jump in circular magnetic structure.

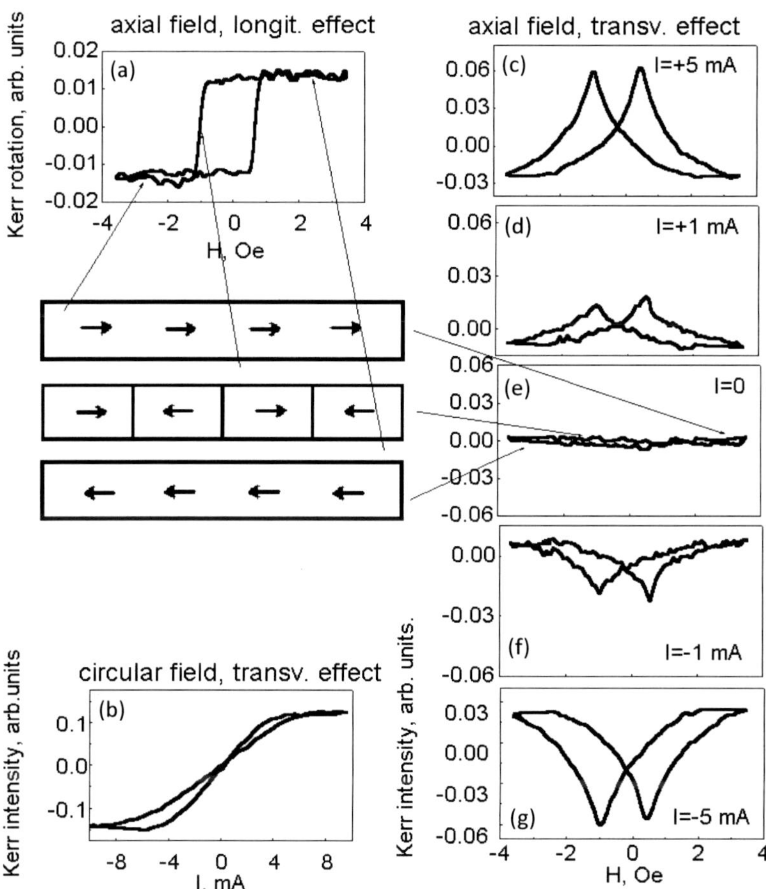

Figure 14.2 Kerr effect loops of Co-rich wire with $x = 0.11$.

The magnetization reversal for the microwire with $x = 0.09$ is close to that of the microwire $x = 0.11$ but some peculiarities are observed. The longitudinal loop obtained in the AC axial field presents a rectangular shape related to the change of the axial magnetization similar to the case of $x = 0.11$. At the same time, the transverse loop has the shape that can be attributed to the successive rotation of magnetization and nucleation of new domains. In spite of this, for the microwire with $x = 0.09$, the magnetostriction constant is negative, and the axial domain structure exists in this wire. Significant contribution of rotation of the magnetization is found in this wire together with domain nucleation, which suggests that this wire occupies intermediate place between wires with $x = 0.07$ and $x = 0.11$.

References

1. Cobeno AF, Zhukov A, de Arellano-Lopez AR, Elias F, Blanco JM, Larin V, and Gonzalez J (1999), *J. Mater. Res.*, **14**, 3775.

2. Humphrey FB, Morí K, Yamasaki J, Kawamura H, Malmhäll R, and Ogasawara I (1987), in (Hernando A., *et al.*, eds.) *Magnetic Properties of Amorphous Metals* (Elsevier, Amsterdam).

3. Chizhik A, Zhukov A, Blanco JM, and Gonzalez J (2004), *Phys. B*, **343**, 374.

4. Chizhik A, Zhukov A, Blanco JM, and Gonzalez J (2002), *J. Magn. Magn. Mater.*, **249**, 27.

5. Chizhik A, Zhukov A, Blanco JM, and Gonzalez J (2003), *J. Magn. Magn. Mater.*, **254–255C**, 188.

Chapter 15

Application of Magneto-Optical Indicator Film Method to Study Domain Magnetic Structure in Microwires

15.1 Introduction

The method of the magnetic indicator films (MOIF) has been recently developed for the visualization of magnetic fields and study of the magnetic flux initially in superconductors [1, 2]. Due to wide experimental possibilities, this method has been applied also in studies of complex magnetic structures. Particularly, the distribution of magnetic moments in a nanocomposite magnetic multilayers consisting of exchange-coupled ferromagnetic films has been successfully studied using this method [3, 4].

It was assumed that MOIF microscopy was applicable only for a different kind of materials with ideally a flat surface, when the indicator film is closely placed to the investigated sample. Here we show that the MOIF technique works well in other type of the samples, that is, when the surface of the investigated material has a significant curvature, as is the case of magnetic wires.

Magnetic Microwires: A Magneto-Optical Study
Alexander Chizhik and Julian Gonzalez
Copyright © 2014 Pan Stanford Publishing Pte. Ltd.
ISBN 978-981-4411-25-7 (Hardcover), 978-981-4411-26-4 (eBook)
www.panstanford.com

15.2 Experiment

The domain magnetic structure of the sample was studied by means of the visualization of the leakage field, which was connected with it and appearing on the surface of wire. For this, the indicator film has been placed directly onto it, as shown in Fig. 15.1. The main element of the indicator is a thin transparent Fe-garnet film containing Bi to produce an increase of the Faraday rotation of the plane polarization of the reflector light. Such film has magnetic anisotropy of the type easy plane, so that in the absence of external magnetic field the magnetization of indicator lies at the film plane. However, normal scattering magnetic field component of the wire forces the magnetization of such indicator film resulting in the local change of the magnetization direction of indicator film in accordance with their value and sign. In the polarizing microscope, the light falls perpendicularly to the indicator film, and passing through it experiencing dual Faraday rotation with the angle of

$$\alpha = k_{\mathrm{f}} \cdot \mathbf{M}_{\mathrm{pr}}, \tag{15.1}$$

where k_{f} is Faraday's constant of the magnetic media and $\mathbf{M}_{\mathrm{pr}}$ is the magnetic moment component parallel to the light raw, since it was reflected from its lower surface of that covered with thin Al by the mirror. If the polarizer and the analyzer of the microscope are slightly uncrossed, then depending on value and sign of stray magnetic fields, i.e., from the degree of the local magnetization of sample the picture of the distribution of the light intensity strictly corresponding to the magnetic structure of sample can be observed. In this case, its image itself looks like variations in the light and dark tones with the intensity proportional to the normal magnetization of the investigated sample.

If the polarizer and the analyzer are crossed, then the picture will be completely different: The light intensity for the domains with the opposite magnetization are identical and then will be visible only transition regions, which correspond to domain walls.

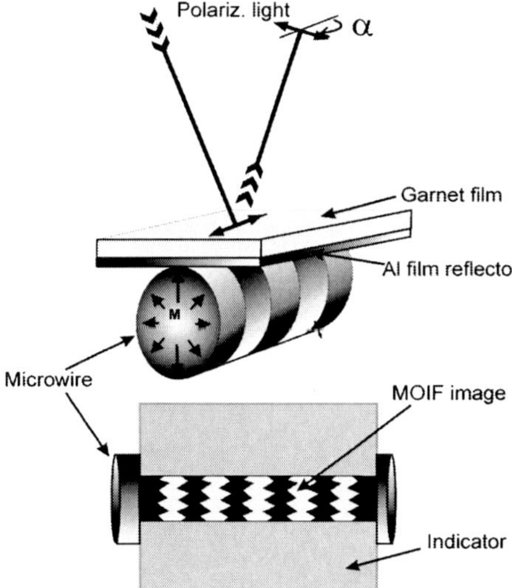

Figure 15.1 Schematic picture showing the experimental setup configuration (a) and the MOIF images from Fe-rich wire (b).

15.3 Experimental Results and Discussion

The fragments of 10 cm-long amorphous wires with 120 mm diameter and the following compositions have been studied: $Fe_{77.5}B_{15}Si_{7.5}$ with positive magnetostriction constant (3×10^{-5}) and $Co_{72.5}B_{15}Si_{12}$ with negative magnetostriction constant (-4.5×10^{-6}).

The magneto-optical picture of the domain magnetic structure of Fe-rich wire in the absence of external fields also with different scales is shown in Fig. 15.2. Particularly, Fig. 15.2 shows the distribution of magnetization in the wire typical for the so-called stripe structure with the "open" domain structure. Band domains (Fig. 15.2b) are strongly twisted and their shape is similar to labyrinth-type domain structure in thin films with bubble domain structure. Figure 15.2c shows the magneto-optical image of the same piece of the wire as in Fig. 15.2b, but with the crossed polarizers. It is evident that the neighbor 180° domains become equally bright, but in this case appears the dark contrast of the transition region between the stripe domains, which corresponds to the position of the domain walls between these domains. Their small width (of the order of micrometers), strong curvature, and intensive uniform contrast within the magnetic domain area indicate that $K_{ind}/2pM_S^2 \gg 1$, where K_{ind} is the magnetoelastic anisotropy induced by the residual internal stresses. The analysis of the image of the distribution of the magnetization of this wire shows that its surface domain structure consists of the stripe domains with the unclosed magnetic flux of those elongated along the circle of wire with the magnetic moment the perpendicular to its surface.

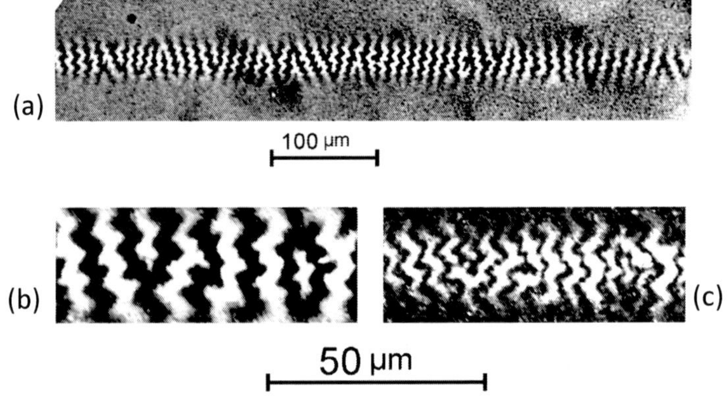

Figure 15.2 Magneto-optical contrast of the magnetization distribution on the surface of the Fe-rich wire (a, b) and (c) MOIF image of the domain walls in the region as indicated in (b).

The character of the observed picture in Fe-rich wires allows us to conclude that the 180° magnetic domains have the stripe structure with the "open" domain structure with the unclosed magnetic flux perpendicular to the wire surface in contrast with previous considerations and modeling of the domain structure of Fe-rich wires assuming this magnetic flux to be parallel to the wire surface [5, 6]. This is, in fact, the first big difference between our experimental results and previous knowledge on this matter.

Figure 15.3a shows the distribution of the stray field arising around the edge of the artificial shallow scratch made on the Co-rich wire surface along its axis (see Fig. 15.3b).

This periodic distribution of the contrast with its inversion in the neighbor region, which we can observe in the photograph, and the absence of any magnetic contrast out of the scratch allows us to assume that the magnetization, at this wire, lies parallel to its surface, with circular orientation and it is formed by 180° domains such as the one shown in the schematic diagram in Fig. 15.3c. Such alignment of the magnetization in the domains creates on the edges of the scratch a stray magnetic field, as shown in Fig. 15.3b.

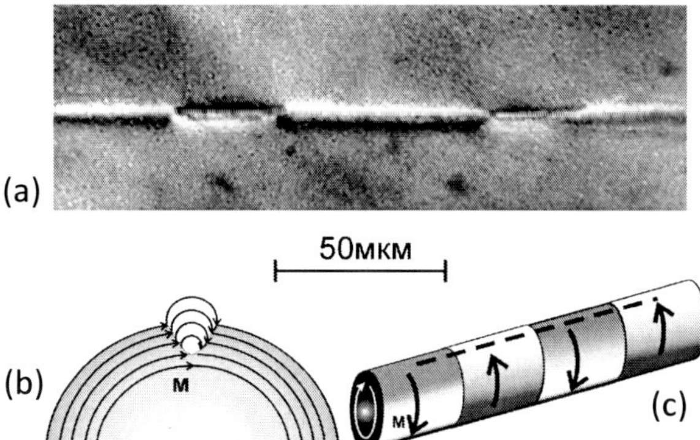

Figure 15.3 (a) MOIF image of the magnetization distribution near the artificial shallow scratch made along the axis of the Co-rich wire, as indicated in (b) and schematic picture of the cross section region near the artificial shallow scratch and the linkage field on the edges of such scratch. (c) Schematic picture of circular domains.

Moreover, a change in the contrast to the opposite in the neighbor sections should be connected with a change in the direction of M upon transfer of one domain to another and, consequently, with the change of the magnetic charge sign on the border of the shallow scratch.

Let us analyze the MOIF images shown in Figs. 15.2 and 15.3. Regarding Fig. 15.2, in which the images of the magnetic field distribution corresponding to the Fe-rich wire are shown, it should be outlined that in the case of the existence of the closure domains on the wire surface with the magnetization orientation parallel to the wire surface, as predicted in commonly accepted models of the domain structure of amorphous wires [6], the MOIF contrast should not exist: In these conditions, the MOIF method does not allow the observation of the contrast from the domains with magnetization parallel to the wire surface. This means that, in fact, the observed domains have the magnetization vector perpendicular to the wire surface.

On the other hand, the image observed in Fig. 15.3 should be attributed to the magnetization parallel to the wire surface. The absence of the contrast from the domain wall between circular domains (marked by the rectangular symbol in Fig. 15.3) should be noted. This fact (excluding the methodological reasons) should be attributed to Néel's wall–type characteristics of such domain walls, in which the magnetization vector rotates in the plane parallel to the wire surface. Therefore, it is reasonable to assume that the domains of the Co-rich wire are rather thin.

The rough domain size can be roughly estimated from Figs. 15.2 and 15.3 (such estimation gives approximate maximum value), being the rough domain size about 5–6 μm for the case of Fe-rich wire. In the case of Co-rich wire, the average domain size is about 50 μm.

Studies of domain structure of amorphous wires with positive (Fe-rich) and negative (Co-rich) magnetostriction performed by the MOIF technique give the following results:

(1) Fe-rich wires possess unclosed 180° surface domain structure with magnetization perpendicular to the wire surface, i.e., without closure domains previously assumed for such materials. Such result can be attributed to the large magnetoelastic anisotropy induced by the fabrication process. Thus, it is known that the magnetic anisotropy energy, W_k, of the closure domains increases with increasing the magnetic anisotropy constant, K, as $W_k = KV$, (where V is the volume of such domain), because the magnetization of the closure domains is perpendicular to the direction of that anisotropy, and at large magnitudes of anisotropy, they do not become favorable, (like in the case of bubble domain structures).

(2) Regarding the Co-rich wires, their domain structure consists of rather big (compared with Fe-rich wires) circular 180° domains, as was assumed from Kerr effect studies.

The application of the MOIF method allows us to revise the commonly accepted interpretation of magnetic domain structures, especially in the case of Fe-rich amorphous wires. Particularly, it should be assumed that Fe-rich wires possess unclosed 180° surface domain structure with magnetization perpendicular to the wire surface, i.e., without closure domains.

References

1. Dorosinskii LA, Indenbom MV, Nikitenko VI, Y, Ossip'yan YA, Polyanskii A, and Vlasko-Vlasov VK (1992), *Phys. C*, **203**, 149.

2. Uspenskaya LS, Kulakov B, and Rakhmanov AL (2004), *Phys. C*, **420**, 136.

3. Kabanov YuP, Gornakov VS, Nikitenko VI, Shapiro AJ, and Shull RD (2002), *J. Appl. Phys.*, **93**, 8244.

4. Nikitenko VI, Gornakov VS, Kabanov YuP, Shapiro AJ, Shull RD, Chien CL, J.S. Jiang, and Bader SD (2003), *J. Magn. Magn. Mater.*, **258**, 19.

5. Tasajo M, Yamasaki J, and Humphrey FB (1999), *IEEE Trans. Magn.*, **35**, 3904.

6. Mohri K, Humphrey FB, Kawashima K, Kimura K, and Muzutani M (1990), *IEEE Trans. Magn.*, **Mag-26,** 1789.

Index